The I in Science

Judith R. Brown

The I in Science

Training to Utilize Subjectivity in Research

SCANDINAVIAN UNIVERSITY PRESS
Oslo – Stockholm – Copenhagen – Boston

Scandinavian University Press (Universitetsforlaget AS)
P.O. Box 2959 Tøyen, N-0608 Oslo, Norway
Fax +47 22 57 53 53

Stockholm office
SCUP, Scandinavian University Press
P.O. Box 3255, S-103 65 Stockholm, Sweden
Fax +46 8 20 99 82

Copenhagen office
Scandinavian University Press AS
P.O. Box 54, DK-1002 København K, Denmark
Fax +45 33 32 05 70

Boston office
Scandinavian University Press North America
875 Massachusetts Ave., Ste. 84, Cambridge MA 02139, USA
Phone +1 617 497 6514
Toll-free 800 498 2877 (US only)
Fax +1 617 354 6875

ISBN 82-00-22661-1

Design: Astrid Elisabeth Jørgensen
Cover illustration: Ferdinand Jendrejewski/NPS
Typeset in 10.5 on 13 point Photina by HS-Repro A/S
Printed on 90 g Partner Offset by HS-Trykk A/S, Norway 1996

To the memory of Nils Magnar Grenstad

Contents

Foreword

Dr. Brown has written a very practical book that is especially valuable for those engaged in the process of training social science researchers. Drawing on material from "The Role of Self in Qualitative Research," a two-week course taught by the author at the University of Oslo, she clearly explicates an educational methodology for the development of self as an instrument in social science research. Those of us who educate researchers need what Dr. Brown provides – a way to teach future researchers how to engage more fully in the research process.

As Dr. Brown asserts, human factors figure in the research process in a number of mutually reinforcing ways. For example, all research occurs in the context of a relationship between the researcher and the participant(s). In this regard the research relationship is essentially a human relationship and implies a mutuality of influence. That is, the researcher affects the social system and the social system affects the researcher. It is Dr. Brown's thesis that we should take seriously what is involved in the human relationships that are the source of our data because they fundamentally affect the research process.

In this book Dr. Brown provides a series of clearly illustrated and well-discussed experiences and exercises that facilitate the development of awareness of self as an instrument in social science research. She draws her pedagogy from the Gestalt therapy of Fritz Perls and the confluent education of George Brown. What I find most remarkable in this approach is the creative integration of content and process, an integration of what is taught and how it is taught. These points are clearly illustrated for the reader. For example, process orientation is taught through process-oriented pedagogy, present-centered contact is taught by means of present-centered contact, and so on.

The book also addresses some of the larger methodological issues in social science research. Most of the current literature in qualitative research methodology speaks of the importance of awareness of self as an instrument in the research process. We know that the personal qualities of the researcher affect the research in a number of ways. However, the current literature is usually silent on how to accomplish this task. The reader is seldom provided a framework or process by which to attend to this critical aspect of the research process. In this regard Dr. Brown fills a gap in our knowledge by providing us with an operationalized educational methodology for developing the self as an instrument in social science research.

From its inception, the research process represents the mobilization of researcher interest and bias toward action. As Dr. Brown states, "The researcher's disposition, inclinations, sensibilities, and expectations all come into play, influenced by psychological factors, cultural background, [and] his or her interests." For example, the multiple choices of an area of study, unit of analysis, conceptual framework, and the formulation of research questions are all based on human bias and interest. The researcher also frames the analysis. In Dr. Brown's words, "People and their interactions are more than a collection of objective measurable facts; they are seen and interpreted through the researcher's *frame* – that is, how she or he organizes the details of an interaction, attributes meaning to them, and decides (consciously or unconsciously) what is important and what is of secondary importance or irrelevant." Each researcher gathers and analyzes information according to her or his personal interests, needs, biases, and blocks. This construction of meaning that takes place in the analysis process is fundamentally a human process. It is also a creative process that is dependent on and limited by the researcher's ability and awareness of self.

Unlike many who advocate the elimination of researcher bias from the research process, Dr. Brown advocates a refreshing approach in which the whole person is returned to the core of the research process. As Dr. Brown states, "Of all the components implicated in conducting high-quality research, the self of the researcher is the foundation on which all else rests." In other words, the key to the integrity of research is the integrity of the researcher.

James E. Barott
University of Utah
May 1995

Preface

The social sciences have a well-developed set of research theories and almost countless numbers of textbooks that give advice on how to perform research. Both the theories and the textbooks deal with subjectivity in research. In the history of science, the dichotomy of objectivity versus subjectivity has been an almost perpetual theme. One of its most significant consequences is the seemingly basic differences between sciences and humanities/ social sciences. In social sciences the focus on objectivity versus subjectivity has resulted in another dichotomy: quantitative versus qualitative research methods.

My personal position is that there are no principal differences between science and humanities/social sciences or between qualitative and quantitative research. Between sciences as between qualitative and quantitative research, there are surface methodological differences, but they are the same in that a person, the researcher, is at the core, thinking and doing.

With this position as the point of departure, I wanted to develop a course at my institute that, in considering the "I" in science, would combine theories and practices in a more sophisticated way than had been done so far. I hoped this initiative would contribute to the development of research methodology in general and particularly in the social sciences.

Judith and George Brown were invited to develop this course. They were known internationally as researchers, scholars, and outstanding Gestalt therapists. This book is one result of the initiative.

Anton Hoëm
Director of Institute for
Educational Research
University of Oslo
May 1995

Acknowledgments

I am especially grateful to the Research Council of Norway for their willingness to support the activities on which this book is based.

The creation of this book was truly a joint effort; my husband, George Brown, was my partner in the task. Together we invented and carried out the teaching that forms the basis of the book. During the writing I depended on him to shore up my sagging confidence at difficult moments, keep me on the path when I drifted off, discuss theory, and remind me to elaborate ("You must say more about this ..."). He read and reread the manuscript at every stage. Astute and penetrating as always, he was altogether indispensable throughout the entire process.

Anton Hoëm's office door, like his mind, has always been open to us. His view is broad enough to embrace our work and to appreciate our addition to his institute. His confidence and optimistic outlook regarding our project kindled my own.

I thank those who read and commented on the first draft of the manuscript and offered specific and useful suggestions: David Winer, my brother, who took time from birdwatching on his vacation; Laurence Iannaccone, who, through his stimulating instruction, encouraged me and drew me into his world of qualitative research when neither of us guessed I would one day ask him to read a manuscript connected to that subject; Marco de Vries, a long-time friend who has inspired and encouraged me, and bravely read early versions of the first chapters as well as what came later; Bjørg Johnsen and Jostein Kleiveland, staunch friends and walking companions with whom I discuss theoretical and practical issues of our work; and Asbjørn Holm, who has responded so positively to our work. Special thanks to Kenwyn Smith for his generosity of time, sincere concern, and specific and valuable suggestions.

Lous Heshusius kindly met with us, sharing her work and her reactions to mine. I am inspired by her provocative ideas and courageous attitude.

Tone Kvernbekk, our student, has become my teacher, recommending books and introducing me to her field, philosophy of science.

I especially thank all the people who stepped into unknown territory and attended our course at the University of Oslo.

Santa Barbara, California
October 1995
Judith R. Brown

Chapter 1

Now the Human Dimension

The human understanding is like a false mirror, which, receiving rays irregularly, distorts and discolors the nature of things by mingling its own nature with it.

Francis Bacon, 1621[1]

This book examines a crucial and neglected aspect of training for researchers – the question of *subjectivity*. The domain of experiential self-knowledge is commonly overlooked because of the bias of the dominant educational philosophy. Engaging the entire person in learning requires a holistic approach, an inclusion of affective considerations with the more traditional cognitive focus. The desired result is a training that encompasses personal experience for people who, as researchers, are inevitably emotionally invested and instrumental in their work. The individual, *the self*,[2] plays a central role in any pursuit of knowledge. Although the recent literature in research methods – particularly in qualitative methodologies – is vast and the need for training for conscious self-involvement is frequently mentioned, any concrete suggestions about how to accomplish this are difficult to find.

This work is intended for instructors, supervisors, directors of instruction, and participants in graduate programs. It attempts to stimulate interest and development in training for personal qualities and those abilities so often mentioned as necessary for the researcher: self-awareness and openness; the ability to be self-reflective, to accept ambiguity, and to think and function holistically. I hope the book serves to inspire and prepare researchers to embody a most important variable in all research – oneself.

A course called "The Role of Self in Qualitative Research," offered to doctoral students at the Institute for Educational Research at the University of Oslo in Norway, is the nucleus around which this book is constructed. It illustrates what is possible in preparing doctoral students to deal with personal issues they face as they embark on and carry out qualitative research studies. The course introduces students to ways of using themselves more fully, consciously, and productively in their work. It has three major focuses: 1) *awareness* to mitigate the likelihood that the researcher's values and emotions will inadvertently, heedlessly, or negligently influence a study; 2) *present-centered contact,*[3] to foster meaningful and productive interactions with individuals, groups, and other sources of data, such as documents; and 3) opportunities to experience *process orientation, dealing with ambiguity, reframing,* and a *holistic outlook.*

The intention of the course is to expand the students' awareness of themselves and help them recognize how their subjectivity can influence their functioning as researchers. By eschewing more familiar cognitive methods and substituting experiences intentionally designed and provided for this purpose, we examine fundamental beliefs and values that can insidiously, and sometimes blatantly, influence the conducting of research. In addition to studying relevant theoretical foundations, we seek ways to increase skills, abilities, and *being-ness.*[4] With an eye toward minimizing error in collecting and working with rich data, we highlight, among other areas,

- the importance of the researcher as one who is *self*-involved in the management of relationships with participators, colleagues, and the researcher's own biographical material;
- questions of reliability and trustworthiness in social science research, especially the impact of the researcher's psychobiography on problem selection and definition, research design, methodology, and analysis of data; and
- extensive laboratory activity in which a variety of experiences are provided to enhance the individual's skills of self-scrutiny, leading to increased self-awareness in a number of contexts.

The origins of the course can be traced to a fortuitous meeting of Anton Hoëm, director of the Institute for Educational Research at the University of Oslo, and George I. Brown, director of the Confluent Education program at the University of California at Santa Barbara. Hoëm recognized that, in conventional higher education, there was no preparation for using the self in research. He was familiar with George Brown's teachings at the University of California and the workshops we (George Brown and I) had led for various groups for more than twenty years in several countries, including Norway. Our background in experience-based confluent education and training in Gestalt therapy for psychotherapists, in addition to the related process awareness training for professionals in non-therapeutic contexts – such as schools, hospitals, and industry – provided a ready and unique source to meet the need. Hoëm wanted to offer his doctoral students similar training to increase their personal awareness and thereby diminish potential distortions in the research process and improve the quality of research at his institute. (Confluent education, the Gestalt approach, and process awareness training will be described and elaborated on in subsequent chapters.)

The two men and I met to create an experiential course in which the students themselves would be the primary subject matter. We decided that the course, "The Role of Self in Qualitative Research," would be an intensive, two-week venture. It would serve to augment the preparation of doctoral students for their research studies at the institute. The first class was held in September 1992. My earliest notes hold the seeds of what evolved into the actual course, and now this book:

- observation – moving from evaluation to description;
- taking a metaposition[5] to enlarge the scope when viewing oneself and others *in action;*
- examining the elements of human relationships in order to improve communications skills and increase acuity when observing interactions of others; and
- understanding the meaning of process orientation through actual participation.

Human factors

Subjectivity is a mental construction that is frequently reified, thought of as a static *thing* that one can manipulate and dole out in varied amounts, as in the following sentence: "It is important to put a limit to subjectivity when using the qualitative approach."[6] Subjectivity is neither good nor bad. It is simply a term that encompasses *human factors,* that is, all that pertains to an individual: thoughts, emotions, feelings; the mind and the senses; the head, heart, body, and soul. As a major theme of inquiry in this work, human factors will be discussed psychologically, sociologically, theoretically, and operationally in connection with the rationale for this book, and related to a general understanding of their place in all cognizance.

Whatever the paradigm governing a research project, whether quantification is a major or minor component or plays no role, whether the instruments employed are the latest scientific technology or simply the researchers themselves, human factors are involved. The people at the helm are motivated by their interest in exploring at the margins of their present understanding. They embrace a particular research philosophy and methodology, choose questions to be investigated, observe processes, collect and analyze data, and attribute meaning in their idiosyncratic way. These realities make dubious the claims of objectivity in scientific investigations in general.

> No longer, in this relational universe, can we study anything as separate from ourselves. Our acts of observation are part of the process that brings forth the manifestation of what we are observing.... John Archibald Wheeler, a noted physicist, states that the ultimate constituent of all there is in the universe is the "ethereal act of observer-participancy." The universe, he says is a participative universe (in Zohar, 1990, 45). We do not, as some have suggested, *create* reality, but we are essential to its coming forth. We *evoke a potential* that is already present. Because things cannot exist as observable phenomena without us in the quantum world, the ideal of scientific objectivity disappears. (Wheatley 1992:36)

The proliferation of the use of qualitative research methodology in fields such as education, psychology, sociology, medicine, and family therapy has brought to the fore the realization that investigator effects must be heeded and made clear if the research is to have worth. A need emerged for changes in graduate education in these fields. This shift in social research has resulted in an outpouring of books related to qualitative inquiry, attesting to the interest in the subject, the need for information, and the complexity of skills necessary for the discovery of in-depth knowledge and understanding of people in various life situations. Authors present guidelines for designing studies, observing and interviewing subjects, interpreting and evaluating data, and even dealing with emotions and fieldwork (Kleinman and Copp 1993). Others divide the field of qualitative inquiry into specialties: different methodologies used for specific areas of understanding and creating knowledge. Handbooks treat all aspects in theoretical and practical detail. The literature often mentions the importance of the subjectivity of researchers; there is broad recognition of the explicit and intimate involvement of the investigator in each stage of a qualitative study. In the words of Berg and Smith, "When we use ourselves as instruments for studying human systems we take on a special responsibility ... to struggle to understand the complex emotional and intellectual forces that influence the conduct of our inquiry" (1988:11). However, their stated need for self-knowledge has rarely, if ever, been explicated or translated into a course including specific training to enhance self-knowledge.

Historical perspective

The scientific community has only recently begun to recognize that the subjectivity of the investigator is a major factor in research – either unintentionally (Rosenthal 1966) or intentionally, as in qualitative methodology. From a historical perspective, one can see the pendulum swinging away from the restricted and restrictive scientific paradigm that evolved in the sixteenth and seventeenth centuries. The view of scientific practice proposed then became the dominant paradigm and has remained so up to the present. Descartes and Bacon are the two names commonly connected with the rise of what became known as the scientific method.

> Descartes showed that mathematics was the epitome of pure reason, the most trustworthy knowledge available. Bacon pointed out that one had to question nature directly by putting it in a position in which it was forced to yield up its answers. The marriage of reason and empiricism, of mathematics and experiment, expresses [a] significant shift in perspective [away from] why things behave to *how*. (Berman 1981:28)

For a quick look at where we have come from in the field of scientific research, we can trace the line back to ancient Greece.[7] Today, when the latest scientific discoveries are revealed in daily newspapers and explained for all to read, it is difficult to imagine ancient times when philosophers attempted to make sense of their world with no texts to guide them and few theories from previous generations. The gods and goddesses of Mt. Olympus were no longer accepted as the causes of all natural phenomena or the passions of human beings. Aristotle, Socrates, Plato, Heraclitus, and Parmenides, among others, are known to us as Greek *philosophers*. They relied on observation, reasoning, measuring, and mathematics to help unlock the mysteries of their world. When they wanted to substantiate that a particular view was the correct one, they found ways to prove they were right, that their interpretation of the universe and humanity constituted the truth. They were personally involved in their investigations and influenced by their biases and prejudices.

Philosophy included under its single banner inquiries into natural phenomena (which we now categorize as the biological and physical sciences) and principles of human nature and conduct (which we call philosophy and psychology). Although science and philosophy are now separate disciplines, vestiges of the old order remain. For example, positions for professors of physics at Oxford and Cambridge have designations such as "The Dr. Lees Professorship of Experimental *Philosophy*" and the "Jacksonian Professorship of Natural and Experimental *Philosophy*" while, at the same time, the chairs of "moral and mental science" are where one finds philosophy professors at Cambridge (Gjertsen 1989:2; italics added). Over time, in Western culture, science has made deep inroads into what philosophers thought was their domain. In actual practice the boundary between science and philosophy has not been, and still is not, always clearly delineated; it is difficult to locate and frequently straddled by those famous in both realms. Among others, Gjertsen mentions the following: Aristotle, Descartes, Pascal, Newton, Bertrand Russell, Whitehead, Wittgenstein, and Polanyi.

The sixteenth and seventeenth centuries spawned many radical changes that made possible the highly technological culture we inhabit today.[8] However, the significance of that period for our present discussion is the pendulum swing from the extreme of personal implication of the alchemist as philosopher/scientist (during the so-called Dark Ages) to the opposite extreme of "pure science," as it came to be known, in which research was to be "untouched by human hands," or unaffected by the humanness of the researcher. The total blending of the alchemist's person with his goal of discovering the philosopher's stone was a striving not only for the transformation of metals, but also, among those who were sincere, for a transformational personal journey from what today we might refer to today as unconsciousness to consciousness, the realization of the higher self. Conversely, with the advent of the scientific paradigm we still labor under today, human factors came to be considered contaminants in the laboratory. The goal was "objectivity": the ideal in physical science, and later in social science.

Throughout the ages methods of determining "the truth" have been of great significance. But it was with Bacon's arrival in the sixteenth century, with his pragmatism and his insistence on testing theories and dealing with technological aspects, that science received its decisive push in the direction of experimental method – thus, the "abandonment" of philosophy for "real science." The division was not only an intellectual one; it might be seen as indicative of the diminishing attention to larger questions of meaning, to being human in the world, and to relationships with others and nature.[9] Perhaps it was then that the socialization of scientists discredited their subjectivity – their values, attitudes, and passions – for fear that these aspects of being human would maximize the possibility of error in their scientific experiments. Was this an attempt to irreconcilably divide science and philosophy? What dangers were latent in this development for a world in which technological goals have eclipsed concern for life and where contaminated waters and foul air are taken for granted? Berman (1981:196) suggests that not until the ideology of Gregory Bateson (1972) was a "fully articulated holistic science available ... that is both scientific and based on unconscious knowing."

Since the beginning of historical records, the human spirit has shown a relentless desire to unlock mysteries, to find solutions to problems and sufferings that are concomitant with existence. Personal agendas are important motivators: for example, a doctor devotes his life to finding a cure for the illness that caused his mother's death; or a woman who grew up in abject poverty makes it her life's work to instigate research projects leading to the creation and implementing of social programs to aid poor mothers and their children. Human factors also include greed and carelessness, and a passion for acquiring fame and glory.

The history of research abounds with examples of studies in which an investigator's intense and unchecked feelings have had unfortunate consequences. This may be obvious. Not as apparent, however, is that emotions are always present, for without curiosity and passionate interest one would not be motivated to learn more or to engage in the search for knowledge. In addition to the usual desires and fears accompanying any endeavor, researchers often have much at stake as they conduct their inquiries. Research demands investments of time, money, and personal reputation, all of which are subject to competition in universities and research institutes. This reality adds to conditions conducive to strong emotions, including desire (perhaps unconscious) on the part of the researcher for specific outcomes leading to financial gain, positive notice, and status.

Science is judged by, among other criteria, the design of the study, the dominant paradigm, the people who do it, and what they may have to gain from their particular findings. Researchers are expected to be always objective, which is understood to mean that research must be conducted under the strictest conditions and untainted by human factors (subjectivity). In reality, while strict laboratory conditions are a laudable goal, scientists can hardly be expected to check their selves at the door. Contrary to a prevailing view of scientists as possessors of incredible virtue, they are not "passionless purveyor[s] of truth.... Committed to truth, unbiased by emotion, open to new ideas, professionally and personally unselfish – the scientist thus described *deserves* sainthood!" (Mahoney 1976:5). In the introduction to his book *Scientist as Subject: The Psychological Imperative,* Mahoney asserts that the scientist has been neglected as a subject of scrutiny, and he questions how one can have confidence in the products of science without a closer look at scientists.

Examples of researchers' falsifying results illustrate the extremes to which scientists have gone to satisfy their needs to prove a particular theory or achieve personal success or satisfaction. Koestler, in *The Case of the Midwife Toad* (1972), tells of Lamarckian scientists who were alleged to have used ink to blacken the pads on the front "hands" of specially bred toads. At the time of the toad controversy, the scientific community was split between those who supported Darwin's view that acquired characteristics could not be inherited and followers of Lamarck, who were out to prove otherwise, in this case by creating false evidence that an acquired characteristic was inherited. Currently, a passionate debate exists among anthropologists regarding the question of "nature or nurture": Are people influenced by inherited characteristics or by socialization in the family and society? In 1986 at a meeting of anthropologists in Seville, a document about violence was drafted. The document affirmed that humans have no innate tendency toward aggression; in other words, aggression is learned. Further, the document accused those who had published an opposing view of "inaccurate work as well as justifying – even encouraging – war." Cultural determinists (those on the "nurture" side of the debate) present at the conference lobbied to have this statement adopted by learned societies. And, indeed, members of the American Anthropological Association were soon asked to vote on the Seville Declaration on Violence. "[The author's] account of this makes chilling reading," wrote Helen Fisher (1994:15) in her review of *The Challenge of Anthropology: Old Encounters and New Excursions,* by Robin Fox. According to Mahoney, idols of science such as Newton, Galileo, and Mendel "doctored" their data to fit their theories (1976:7). Unimpassioned consideration, including good sense and clear perceptions, does not have a chance when one ignores, or is ignorant of, one's fears and desires. In the face of denial, the fears and desires gain the upper hand, and one is steered by a set of beliefs and prejudices and/or motivated by personal gain.

Now the pendulum swings back, not to alchemy, but to the acknowledgment that white coats and protective gloves do not prevent the human factor from influencing a research study. Despite the express goal of objectivity, science has always been very much shaped by the human touch.

Researchers themselves have been the subjects of investigations, although rarely (Mahoney 1976). Robert Rosenthal, a Harvard University behavioral researcher using quantitative methodology, was a researcher of researchers. He realized the necessity for research on the unintentional influence of researchers on subjects. He examined "experimenter effect" in the physical and biological sciences, that is, whether the investigator's observation and examination influences the subjects of an investigation and, if so, in what ways. He questioned whether some of the complexity of the research subject "resides not in the subject himself but rather in the particular experimenter and in the interaction between subject and experimenter" (Rosenthal 1966:vii). His research presents evidence that "the opinions, outlooks, expectations, and theoretical as well as practical biases of an experimenter [or] interviewer, ... even if never made explicit, have a definite effect on the performance or his subjects, whether they are rats or humans" (Watzlawick et al. 1974:112).

Mitroff, who also researched scientists, wrote that "for too long one of the myths we have lived with is that science is a passionless enterprise performed by passionless men, and that it *has* to be if it is to be objective" (1974:23). Scientists themselves have considered this viewpoint to be "simple-minded nonsense" (p. 64). "Perhaps the passionate, and often even irrational, adherence to ideas was the norm or distinguishing mark of the creative scientist" (p. 6). On the other hand, Mitroff asserts, "the existence of such extreme emotions does not prevent scientists from doing excellent scientific work and being reputable and scrupulous" (p. 99).

There is, in a word, no such thing as impersonal science. Humankind is permeated with interest, curiosity, and the unceasing desire to find out about the universe.

A new paradigm

The preceding overview, an obvious oversimplification of 2,500 years of inquiry into the existence of God, the universe, life, and ethics, provides a context and perspective for the incipient movement away from thinking that the methods that have become known as objective and scientific are the only possibilities for gaining knowledge.

Scientific investigations lead to changes not only in our world and how we live in it, but also in theories and further research. Ironically, changes introduced in the scientific community have often met stubborn resistance from other scientists, especially if these changes threatened a prevailing paradigm. A paradigm is more than a simple scientific theory; it is an encompassing set of accepted beliefs that opens the gateways of science to a new world of discovery. "As a system of beliefs and practical guidelines, [a paradigm] endorses an *ontology* (theory of reality; the nature of things) and an *epistemology* (theory of knowledge; the methods of appropriate inquiry)" (Mahoney 1976). Thomas Kuhn (1970) elucidated the process of creation and destruction of scientific paradigms. Darwin's theory of natural selection provides a well-known example of a paradigm that shook the scientific world and still sparks antagonism. A paradigm imposes a set of beliefs on those who accept it, so any allegiance to a paradigm may result in narrowing the recognition, even the consideration, of possible alternatives. Deeply internalized paradigms become dogma, taken-for-granted information that scientists come to think of as absolute truth. Although these scientists may not be aware of their unquestioning loyalty, the paradigm's influence remains. A new paradigm may be a real eye-opener; with time, it can become a mind-closer.

It has taken time for qualitative research to gain some acceptance in academia. The social sciences have struggled to adapt to models developed for research in the physical sciences, while trying to deny the poor fit. Whereas models for physical science research may successfully attempt to isolate aspects of a phenomenon and examine parts, this is not feasible when dealing with individuals. For example, to discover the best conditions for growing roses, all variables would be tested, one at a time, or in specific combinations: climate, soil, fertilizer, amount of sun, water, and so on. However, as Spiro (1986) points out, in researching the best treatment for a particular illness, dealing with parts will not work. Patient responses depend on many qualitative factors. For example, does the patient have faith in God? Faith in the doctor? Does the doctor believe in the treatment? What emotional factors are influencing the general mental and physical health of the patient? These variables, and many more that might influence the outcome of a study, are impossible to isolate.

Caroline Wiener (1990) speaks to the topic of research that would "capture the complexity and creativity of meaningful human activity." She maintains that, in describing research as "qualitative" or "quantitative," we "have borrowed from the physical sciences to our detriment." She continues: "Social science lacks a good word for what qualitative work is all about. Or perhaps ... we have too many words for it: symbolic interactionism, phenomenology, ethnography, action theory, ethnomethodology, participant observation, historical sociology, conversational analysis, social linguistics, hermeneutics and this does not cover them all" (p. 11).

Whatever the distinctions made among these categories, the thread that connects them seems to be the necessity for strategies to study people in diverse natural settings where researchers are personally involved as their own instrument – in interviewing, observing, studying documents, analyzing data, and creating a narrative to present to the world. It is time to look at training that would make it possible to enhance their effectiveness in this new mode of inquiry. We need to design educational approaches to training in the use of subjectivity.

The structure of the book

The above has introduced the impetus for this book and offered examples of the use and misuse of subjectivity. It has presented the historical context for the current need for experiential subjective education in training graduate students to do research. The definition of "acceptable science" has broadened to include new approaches in the social sciences. The "I," the personal component of the researcher, can no longer be neglected in educating researchers for the ever-expanding world of scientific investigations.

In Chapter 2 the focus narrows from the general picture of science to the individual investigator, who is the conduit through which all data flows in qualitative inquiry. Unlike water pipes, which do not mix with the substance flowing through them, the person conducting the inquiry is, and needs to be, permeable – involved, influencing, and influenced by the participators. Thus, self-scrutiny becomes a primary obligation of the researcher. In addition, complexities of interaction call for certain distinctive qualities and abilities. These demanding expectations that researchers face warrant specially designed education.

Chapter 3 presents confluent education as a method that combines the affective and cognitive realms of teaching and learning. This is followed by principles of the Gestalt approach. Both have direct application to subjectivity-centered education as well as close affinity with social science inquiry – non-linear, paradoxical, process oriented – and are foundational for the training undertaken in the course described.

Chapter 4 provides an overview of fundamentals of communication, metaprocessing, and metacommunication, as well as their relevance to subjectivity and research. It constitutes a basis for understanding the practical course work that follows.

Chapters 5 and 6 present the concepts, structure, content, and methodology of the course offered to doctoral students at the Institute for Educational Research at the University of Oslo. It also describes the pre-planned exercises and much of the interaction between members of the class. Aspects of the course are treated in detail: Gestalt group experiences, student interaction, and process awareness work, each accompanied by a rationale and student reactions.

Chapter 7 includes excerpts from letters written by students who have attended the course, illustrating some of the effects of the two weeks of training. The letters report how the students apply what they learn to help meet the challenges of their research studies.

Chapter 8 sums up this work through reflection and retrospection using a metaperspective.

Chapter 2

Subjectivity and Relationship in Research

In contrast to objects ... human relationships are not phenomena which exist objectively, in their own right, as it were, nor is it possible to have a consensus about their properties.

P. Watzlawick (1990:15)

All research is participatory. I have discussed in Chapter 1 how subjectivity – the person of the researcher – is inevitably involved in research studies, contrary to the predominant idea that a researcher must be divorced from the subject of inquiry in order to keep the data "pure" from personal bias. As they work scientists naturally experience a range of emotions and thoughts, some of which threaten to bias, distract, and even disable them. The following was written by a student doing research in a classroom setting:

> I'm watching the children! I'm watching the teacher! There are so many things going on at the same time! My head is spinning! What should I write in my log? What should I leave out? And to top it all it's just impossible. The minute I put my head down to write, I'm disconnected from what is happening. And if I tune in to what I hear, I may lose an important thought. They didn't tell me that this would be so apparently chaotic and so unnerving. If I'm the instrument, I need to be sent to the repair shop. (Ely 1991:48)[1]

These human factors cannot and should not be denied:

> The distinctive feature of the knowledge [personal knowing of the professional] is that it is interactive and contingent on the participation of both the professional and the participator. The fact that the practitioner

> is always a participant observer makes it untenable to maintain the assumptions of a detached observer as in the traditional research setting. (Hoshmand 1994:13)[2]

Interactions

Research that includes interaction between a researcher and human subjects is fraught with complexity: inevitably there is entanglement between the knower and the known. Just as witnesses to an event tell widely differing stories of what happened, researchers gather diverse information, according to their personal interests, needs, biases, and blocks. People and their interactions are more than a collection of objective, measurable facts; they are seen and interpreted through the researcher's *frame* – that is, how she or he organizes the details of an interaction, attributes meaning to them, and decides (consciously or unconsciously) what is important and what is of secondary importance or irrelevant. The researcher's disposition, inclinations, sensibilities, and expectations all come into play, influenced by psychological factors, cultural background, her or his interests, and the relationship between the researcher and the participator(s) in the investigation.[3] The following remarks from a researcher highlight these ideas:

> What is integrated in me from the course is 1) an awareness of how different people pick up different things in the same situation, and the degree of influence that makes to the data collected. That makes me underline still more that open questions in interviews and observations have a personal color, and that this color is explicit for the researcher, and has to be made explicit for the reader of the document. 2) The understanding that methodological questions also consist of working with oneself as a person.[4]

When investigators encounter participators, a communicative event takes place. The investigators are probably never completely detached or dispassionate as they interact with others in face-to-face contact. They hear, smell, touch, and see the people before them; they take in their bodies, eyes, and facial expressions, and they have reactions. The particular characteristics of a participator evoke responses, and the totality of the person awakes in the researcher a sense of him- or herself *in relation* to this person. Even when learning about people through their writing, or through the writing of others, or voices over the telephone, researchers tend to respond in some personal way, size them up, categorize them, and know how they feel about them. It is difficult, if not impossible, to remain exclusively "analytical, distant, and asocial," as Lowman (1988:174) described the "Investigative" type (Holland 1973), even if the researcher might prefer to be so. Their reactions may be guarded or open, vague or intense; either way, they do respond. One's responses depend on the meaning one makes.

Creating meaning

Meaning is not inherent in a situation, an object, or a person. As researchers take in information through their senses, they organize and make meaning out of it with their minds. For example, suppose a researcher has been invited to meet a participator in his research project. When the researcher arrives, the participator answers his door with a gun in his hand. What does this mean? How does the researcher respond? Of course, it depends on the context. The situation helps determine how one organizes what one sees, hears, and otherwise senses.

In addition to the situation, one determines meaning by what one "knows" from a lifetime of experiences. An interviewee might say something or behave in a way that causes an investigator to succumb to emotions that have little or nothing to do with the situation at hand but much to do with personal history. In such cases investigators, responding with anxiety, anger, or an excess of empathy, will lose clarity of perception. They may lose the ability to project themselves into the others' situations to imagine how they feel; there remains little space for connection. They are likely to skew information to justify the "correctness" of their reactions; thus, the collection and interpretation of data lose reliability. This researcher tells how she struggled with the problem:

> When doing interviews, it was surprising to me when the interviewee said something that struck home. It is at these points I was particularly pleased to have a tape recorder. When a point hits home, it is often hard to concentrate on what's being said ... [and] not to suddenly jump in and tell the person being interviewed everything about myself and that I felt exactly the same. Emotions play a big role in this type of research. (Ely 1991:61)

An investigator whom I interviewed gave the following account of her attempt and failure to remain distant and asocial in an interview she was conducting:

> We were working on this study of "teacher thinking." Each participator watched the same video involving a teacher in a classroom and then would answer questions about it. The aim was to give every person who was participating the same interview so that we could compare people within the group, and across groups. If we shifted what kinds of questions we asked, or if we went into depth in any kind of questions, we would be altering the data we got and therefore invalidating that study. I was an interviewer in [this study] and I had a dreadful time not responding to the person I was interviewing. I'll give you an example. This one sticks in my mind. This young woman watched the piece of videotape, just like everyone else did, and one of the things she noticed in the videotape was that the teacher snapped her fingers at the children like this [she demonstrates] and said, "Come on, let's go." When the participant talked about that, her voice started to choke up. She was very upset. I saw her face was getting red, her voice was changing, and being in that interview situation my job was not to respond, not to say, "What's happening to you?" which is what I really wanted to know. It was obviously a deep experience. Well, my struggle right there was to say something but I had to *not deal with it* [italics added]. But when she started to talk about it a second time, I needed to know what was happening. I couldn't hold back; I went with my Gestalt training: "I hear your voice is changing." I asked and she was able to tell me.
>
> The situation had brought to her mind a time when she was a child and her teacher had snapped her fingers at her. She was learning English as a second language; she was paying attention but she didn't have the language; she couldn't respond. And it was terrifying for her. The teacher wasn't aware of anything that was going on for her as a student. And here in the moment of the interview she's reexperiencing that. At that point there were tears in her eyes, and she was really surprised at herself, that here had been such a buried memory that was suddenly recalled. I'm dealing with this in terms of my own interviewer

role saying, Here I have some data which is emotionally full of impact for this person; it's meaningful for her, and if I didn't use that technique, I wouldn't have gotten it. If I didn't recognize her in the present, I wouldn't have gotten the story. And the story was meaningful to my data, as well.

When we were finished with the interview, I asked if she wanted to talk anymore about the incident, if she needed some time for closure. She said she was really surprised at herself, that it was so powerful. That's some data for me, on the power of this, of using yourself as an instrument, being aware of who you're talking to and what's going on for that person, and not taking a back seat, a sterile position. That is like having a machine ask questions and the interviewee respond.

Self-scrutiny

The above excerpt exemplifies subjectivity, the use of all of oneself as investigator, evidenced by such phrases as, "I needed to know," "my struggle right there," and "I couldn't hold back." In addition to being aware of her feeling responses, the interviewer used her Gestalt training "to recognize [the interviewee] in the present." She also commented on what she heard; the statement "I hear your voice is changing" served as an invitation for more information.

If the participatory, interactive method of acquiring knowledge is to be advantageous, the process of engagement must be acknowledged, noted, and reflected upon; the researcher collects and considers this material, not just one time at the beginning of a study, but repeatedly throughout the course of it. Consequently, investigators must step into their own spotlights and with a cold eye assess their behavior and thoughts: their motivations in choosing a research topic, the effect their research has on them, and their interactions with colleagues and participators and with the data they collect. Some aspects of the self are easier to detect than others. The following are some reflexive questions one researcher keeps in mind:

In the journal I try to be as frank as possible. What do I think of the interview situation? Boring? Interesting? Nice? Did I like myself in the situation? Did I like the other one? What about dislikes? How was the general

> atmosphere and how did it develop? How was the other dressed? How did he or she talk? How about my dress and language? Did I ask leading questions? Was I open and confident in the situation? Did I have a hidden agenda? Did I try to control the situation in some way? Did the other try to manipulate me some way or other?[5]

Many social scientists have suggested that the role of the researcher needs to be examined.[6] Berg and Smith (1988) are among those who have specifically highlighted the place of the researcher in social investigation. They ask, Where do we (social scientists) "position our selves" in connection with our research? They emphasize the serious responsibility that qualitative researchers have to science – and to the subjects of their studies – to examine minutely their relationship with their research (p. 11). Their book explores "the emotional and intellectual struggles that attend researchers as they formulate ideas, collect and interpret data, and build theory" (p. 11).

Students who begin research in the social sciences may not expect to embark on an odyssey of self-discovery. An investigation directed inward toward the self of the researcher may at first glance seem a curious and questionable distraction from the "real" investigation, that of the individual or groups who are the focus of the research study. Yet, since all information goes through investigators during their collection of data and their subsequent analysis and write-up of the material, they need to be well acquainted with who they are and how they use themselves. Those with experience and expertise in the field see self-discovery as essential. Twenty-five years ago "self-analysis" was listed as a topic in participant observation with importance equal to "observation, informant interviewing, document analysis, respondent interviewing" (McCall and Simmons 1969:preface). Another author regards awareness of "biases, blind spots, and cognitive limitations ... as high a priority as theoretical knowledge" (Simmons 1988:303). Simmons recommends psychotherapy as "a routine and essential part of graduate training" (p. 303). In addition to the question of practicality, this also raises the question whether meaning gained in one context – the therapeutic – would transfer to another – the research role. The researcher's inquiry, in other words, must at all times be bifurcated; one investigation is directed outward, the other inward to the self. (For discussion purposes these investigations are delineated as separate. In reality they are not.)

Although the injunction "know thyself" is commonly accepted as good advice, where or when in our development do we actually learn how to increase self-awareness?

The learning we acquire in our homes and schools generally has a focus outside ourselves. We learn to manipulate our environment to minimize anxiety and, at the same time, get from our environment what we need to grow and develop.[7] We learn subject matter, but we ourselves are not the subject matter. Family members and teachers may seem to be informing us about ourselves – what kind of people we are and what is right and wrong about us, when we merit praise and when we deserve criticism. This kind of "teaching" actually constitutes repeated attempts to mold and control. It distracts us from consciousness of our moment-by-moment existence, thereby discouraging any inward-directed, *intra*personal process focus. Indeed, it encourages us to disregard our experiences and pay attention to the reactions of others.

While still young we internalize our observers and judges, adopting their attitudes, criticisms, and wishes as our own. Thus, we learn to adapt to our family, community, and larger culture. We become socialized. We tend quite naturally, then, to become *other*-directed, disconnected to some degree from ourselves, detached from our emotions, feelings, desires, fears, and even thoughts. Justifying our actions becomes more important than awareness of our intentions or motivations; reasoning *why* we feel something becomes more important than experiencing *what* we feel; controlling excitement takes precedence over acknowledging our emotions. We shove self-awareness into the back seat as we split ourselves off and become "the way we should be." Basic awareness is rarely, if ever, called to our attention.

I excerpt the following from my recent interview with an American doctoral student. I mentioned the course we were teaching to increase self-awareness, "The Role of Self in Qualitative Research":

> Author: I wonder what your reaction is when I talk about self-awareness.
> Student: I'm really bound in "social construction." Everything, even if it's me, myself, it's socially constituted. I would have to learn how to interact with myself, and the kinds of questions to ask myself in order to do that.
> Author: Would you know how to do that? To know your part in an interaction, how you are influencing the other, how are they responding to you? Is there a possibility that during an interaction you could somehow step outside for a moment to discover you?
> Student: No. For me that's the incredibly difficult part. At this point I don't have any idea. I have had no learning experience like that at all. I don't think there's a lot of work out there, that I've come across anyway, that helps with that. The only thing is on taking field notes. The difficulty for me is that has to be in the moment. And if you're taking field notes and trying to do all this, that in the moment is really difficult. Yeah, how do you do both? And is it possible to do both? It might help for me to have questions to ask and things to be aware of.

Our work with graduate students and other populations has taught my colleagues and me that this lack of experience in self-awareness is not unusual. And although many people recognize the importance of self-awareness for the researcher, few know an actual approach to the achievement of self-scrutiny. Many perceive self-awareness as somewhat dangerous, requiring (as it does) a willingness to free oneself of self-concepts, self-deception, and long-held beliefs about oneself and others. It requires a willingness to use the same interest and curiosity to experience oneself as one does in researching others.

People often have little idea of what they are doing or experiencing. When first viewing themselves on videotape, many are surprised at what they see and hear: "I didn't know I do that with my mouth. Look how I'm moving it all the time." "My voice is so soft. I thought I was speaking so loud." "I thought I was quite relaxed, but I look really nervous." At a deeper level people's intentions and motivations frequently mystify them. Any experience of inner reality may wholly evade them as they explain away their feelings, reactions, and behavior to themselves and others in habitual ways that they hope make sense. The following comment by an acquaintance exemplifies this pervasive maladaptation: "When my father died I really knew what I was feeling. I think my grief over losing him was the first pure emotion I have felt since I was three."

The open perspective

The qualitative researcher must have the capability of being open to people and new situations and ideas. Trusting oneself to be able to handle whatever comes along facilitates openness. This implies feeling secure about one's emotional reactions and being able to tolerate emotions expressed by others. The abilities to identify with another's experience – even when that experience is different from one's own – and to get beyond judgments of inferiority and superiority involve temporarily dispensing with guardedness and moving with the process of the interaction, engaging in the natural unfolding of a sequence of events. It means coming, not to *the* truth, but to the truth of oneself, the other person, and the *relationship* in the moment. This excerpt, from a doctor who attended our course, speaks directly to these issues:

> In our interviews we try to get a picture of how it is to be a doctor or specially trained ambulance crew called to an unconscious person where you as a rescuer within minutes have to decide if you should resuscitate the patient or not. We also ask them questions about how it really is to "decide life over death" and how it is to live and work with acute medicine out of hospital.... I am all the time very aware of my own feelings and reactions to the things said and done by the interviewed person.... And I tried to use "the non-verbal communication" with the subject of the interview. (Examples: "I can hear that your mouth is very dry now." "Your voice is so low/sad when you tell me this.") I could also tell them what it did to me to listen to them and to go in to the things real deep together with them. This helped me build a kind of confidence/honesty in the interview situation; I think it made me less dangerous as an interviewer. I was open and honest so they could be open and honest. If we got into difficult situations, I could solve the problem because I had learned how to put words on what I felt and thought and what I saw.[8]

Interviews encompass more than asking and answering questions. Watzlawick gives an example of the nonsense spoken when a "normal volunteer subject" feels required to answer a question he apparently prefers not to answer. The following is a portion of the reply (almost a page long) to the question "How does it work out, Mr. R., with your parents living in the same town with you and your family?"

> Well, we try, uh, very personally I mean ... uh, I prefer that Mary [his wife] takes the lead with them, rather than my taking the lead or what. I like to see them, but I don't try too much to make it a point to be running over and have them ... they know very definitely that ... (Watzlawick et al. 1967:77)

Spontaneous, unpremeditated responses rather than deliberate, formularized techniques can create an opening to a real meeting between two people. In the interview relationship, the researcher is the key to such an opening. In her article "Freeing ourselves from objectivity: managing subjectivity or turning toward a participatory mode of consciousness?" Lous Heshusius raises several questions relevant to researchers, all dealing with the complexity of knowing and understanding between people. Basic to any interview is grasping the *meaning* of the participator – not only the words spoken, but also the totality of the speaker's experience. Accomplishing this aim depends, in large part, on the attitude of the inquirer, which must be one of openness to this other person's total experience. Heshusius relates a fictional account (from Naylor 1989) of a young man who returns to his island homeland to conduct research,

> hauling himself back from one of those fancy colleges mainside, dragging his notebooks and tape recorder.... And then when he went around asking us about 18 & 23, there weren't nothing to do but take pity on him as he rattled on about "ethnography," "unique speech patterns," "cultural preservation," and whatever else he seemed to be getting so much pleasure out of while talking into his little gray machine. (Heshusius 1994:15)

Heshusius asks what this researcher, who totally missed the mark in gaining any meaningful information, would have needed to "come to a fuller understanding" of these people whom he interviewed. She rejects the idea of "managing one's subjectivity," of partitioning ourselves such that one "part" could do something to another "part" – in this case subjectivity, an abstraction, an idea created in the mind without actual existence. She equally rejects the notion of "accounting for" one's subjectivity. Regardless of how the interviewer in the story might analyze his confessed biases and prejudices, he had approached his subjects full of certainties about them and their existence and had come away none the wiser; rather, he had simply reinforced his prejudices. Heshusius argues that what is required is "participatory consciousness, ... the awareness of a deeper level of kinship between the knower and the known." Quoting Schachtel (1959:225), she asserts that this deeper level of consciousness engages "the participation of the total person" and "requires an attitude of profound openness and receptivity." It involves (she refers to Schachtel 1959:181) "a temporary eclipse of all the perceiver's egocentric thoughts and strivings, of all preoccupations with the self, and self esteem" (Heshusius 1994:16). Paradoxically, in this state of boundaryless meeting, one forgets oneself but does not lose oneself.

Self-scrutiny is not just close observation of oneself, which suggests a division of the person into observer and observed, but more a mindfulness, an awareness of one's moment-by-moment processes through *direct experience* and largely physical rather than cerebral discovery. The "I" includes physical sensations, thoughts, emotions, and all that is beyond one's consciousness – sometimes called the unconscious, the subconscious, or the psyche. Constant scrutiny could become counterproductive; it would interfere with contact with others, spontaneity, and joy in life. It would be a full-time job and, one would think, wholly unrewarding. Yet paying attention in the moment, a fleeting tuning-in *without judgment or evaluation,* can foster self-awareness and a heightened awareness of the process of interaction, which are not only useful but necessary in the research process.

Interactions in relationships

Interaction, by definition, involves more than one person. When two or more people come together in any kind of association, they communicate; they engage in "sequences of human interaction that are strictly governed by a complex body of rules" (Watzlawick 1967:42). What are these strict rules in the case of the qualitative investigator? Who makes them? Who breaks them? Are they decided before or during the interview? Do the same rules apply to both parties? Are they implicit or explicit? Is the interviewer aware of some or all of them? Can the interviewer change the rules? If so, what happens then?

Amid all the mystery of communication between two (or more) people, there remains one certainty: mutual influence. Recent research in physics has revealed many surprises that highlight problems of mutual influence between the researcher and the researched, even when the researched is not a person but a particle. What is more, it seems "experimental intention, not just the action of the experimenter, can influence the experimental outcome" (Miller 1993:71). Regardless of the research done or who is doing it, the very fact that the researcher (or technician or other staff member) is human has an impact on the research, just as the research affects the researcher. When the "pigeon lady" handled Skinner's birds with loving care, what impact did her affection have on these subjects of his behavior modification research?[9] The existence of the investigator–investigated relationship and its effects must be made explicit and considered and noted by the researcher.

In qualitative inquiry the ways in which the researcher and the participator influence one another constitute a complicated and ongoing concern. Berg and Smith wrote that "all social research occurs in the context of a relationship.... The self-scrutiny process is difficult and complex precisely because *both* the researcher and the 'researched' are simultaneously influencing each other" (1988:31). Interacting with others in a research context, whether with an individual, group, or organization, requires a willingness to experience what is occurring at several levels: individual, interactional, verbal, emotional, and psychological. The emotional level relates to strong feeling responses in the moment; the psychological relates to mental phenomena such as attitude, motivation, behavior, and characterological qualities and traits. Psychological and emotional factors are powerful and may facilitate or hinder our abilities to be authentic, open, and receptive to others.

When participators sense that an observer likes and approves of them and is interested in their well-being, they respond accordingly. When they sense a negatively judgmental observer, they alter their behavior, even though they might attempt to disguise their authentic reaction. A relationship is created through verbal and non-verbal communication. Since all behavior is communication, it follows that we cannot *not* communicate. When in the role of observer, one's behavior reflects more than one might realize. The observer's responses may be as subtle as an intake of breath, the slight raising of an eyebrow, or a change of voice, but astute interviewees detect them and interpret – or misinterpret – their meaning. We send out signals when we want something from, or take a dislike to, the one we observe, even when we are unconscious of our feelings. We give ourselves away, without realizing it.

Information important to a study is not only related to explicit interactions and relationships; it is inherent in the relationships themselves. That is, the verbal meaning of communications accounts for but a fraction of the information. Relationships, the contexts that give meaning to verbal communication, are expressed in innumerable ways, including who stands and who sits during an interaction, who speaks first, who asks questions, who interrupts, and various other forms of non-verbal communication. In a word, interaction is difficult, if not impossible, to capture, unravel, interpret, and analyze completely and conclusively. While investigators continue to research interaction, the bits and pieces they discover about the mind, the brain, chromosomes, and communication cannot reveal all of the mysteries of human intercourse.

The effects of an investigator on a study are always potentially powerful, and they can alter a process in ways that were not intended and not in the design of the research. There may be no definitive findings to support the inevitability of researcher effect in the natural sciences. However, in the social sciences, since the influencing presence of an investigator is assumed, the trustworthiness, usefulness, and value of a study all depend on the researcher. This fact requires serious consideration of what can be done to prepare for this role.

Enhancing the researcher as instrument

To this point I have emphasized the necessity for the researcher to be capable of using herself or himself well as an instrument. Self-awareness and conscious, responsive interaction require instruction and practice quite different from that traditionally offered to graduate students in university. In addition to personal qualities already mentioned (openness, self-trust, security with one's emotional reactions, ability to identify with another, getting beyond judgments, and moving with process), the skills and qualities discussed below help to sharpen the instrument and make it more effective in utilizing subjectivity intentionally and consciously in research settings. I will also describe actual activities from our course as examples of such education.

Consider this analogy from outside the realm of science. A professor of music is lecturing on the minuet. He describes how it was once used solely for dancing and then became integrated into concert music for listening. He plays tapes to demonstrate how different composers used the minuet form and variations of it, and what might be referred to as the "left brain" details, engaging the intellectual faculties of the audience. Then he changes his tack and calls upon his students to respond in quite another way as they listen to music:

> You're not supposed to sit there and let this stuff wash by. The audience is part of the performance. You must somehow use your *body* while you're listening. You must somehow activate your *intellect* and your *emotions* when you listen. In doing so you'll become part of that great and holy triangle of composer, performer and audience. Every part of

> that triangle is necessary. Without the performer, the composer's music is merely dots on the page. Without the audience the performer is simply playing to the ether with no impact, and that wonderful circle, folks, that cycle, is what drives us all in a musical situation. So the audience is part of the performance, and being part of the performance means encountering the *music in as many ways as you can at as many levels as you can* [italics added]. (Greenberg 1993:Lecture #21)

The same may be said of the researcher who actively receives information from an interviewee. The intellect *and* the emotions *and* the body must all be activated. Specific qualities and capabilities are mentioned in the literature: Ely lists the following characteristics as "important in order to meet the rigorous demands of this [qualitative] research: flexibility, humor, accepting ambiguity, accepting one's emotions, ... a balance between empathy and the pursuit of a distanced, non-judgmental stance" (1991: 113). Other lists are similar.[10] The course work, as developed and described in this book, stimulates and intensifies those everyday experiences that we normally acknowledge fleetingly or ignore altogether. For it is in our normal process of living that we either invest ourselves and use the full spectrum of our being for creative engagement, or simply skate over the surface of life.

The task of studying or treating individuals, groups, or organizations – each of which is a constantly changing field – requires *presence* and *responsiveness* of the investigator. This truly creative process calls for a fluid and dynamic approach: "The researcher is the instrument of qualitative inquiry ... the quality of the research depends heavily on the quality of that human being. Creativity is one of those special human qualities that plays an important part in qualitative analysis, interpretation, and reporting" (Patton 1990:433).

Openness is commonly mentioned as being important for any successful person-to-person interaction. Participatory consciousness demands it.[11] Openness presumes one is willing to know others and be known by them. The word itself suggests a fluid and dynamic approach to life. One imagines the inhaling of full breaths, a body that moves with flexibility, an expressive face. It brings to mind personal strength and "systemic coherence":

> The ability to rally from setback is one of the main characteristics of a system with high ego-strength or systemic coherence.... The capacity not to give up under pressure is therefore vital.... Ego strength involves not a rigid unyielding ego, but rather one that is open and accepting, capable of becoming disorganized without falling apart. This means having enough to risk being shocked, or at least exposed to phenomena that are unexpected, unsettling, and contrary without feeling the need to control them. (Montuori 1992:196)

There may be no solely cognitive way to teach what Heshusius referred to as an "attitude of profound openness and receptivity," or to teach abilities and skills necessary to break down the walls we build between ourselves and others (1994:16). There are no prescriptions or rules to memorize for questions that begin, What should I do when ...? However, in a safe and cohesive group, one can *experience* how one holds back, remaining remote and unavailable for participation in any moment. One can experience how "trusting a process" feels and, on the other hand, how one may want to control a process for a particular outcome. Experiential exploration of three fundamental skills – seeing, hearing, and giving personal response – can locate blocks, interruptions, and interferences of process. This kind of experiential education attempts to increase personal and interpersonal development in order to realize and activate one's full potential.[12]

Most of us take *seeing* so much for granted that we usually pay no attention to how we see. Needleman refers to "the revolutionary act of looking attentively at what is right before one's eyes" (1985:10). We frequently and unknowingly substitute imagining for seeing. Perhaps we do this to avoid discomfort – by replacing what is actually there to see with what we expect, wish, or think should be there. When we look at something or someone, we tend not to question our interpretation; we suppose that what we see is what is really there and that everyone else sees the same thing.

Hearing and getting the intended meaning are especially troublesome for many reasons. Some of the more obvious confusions stem from uncertain meanings of words chosen to express thoughts and experiences; misleading abstractions and metaphors; tone of voice; stress on a particular syllable or word; and suggestive facial expressions and gestures that are not congruent with the verbal message, resulting in mixed messages. The context, upon which meaning always depends, may be unclear, adding to the uncertainty or obscurity of a message. Sometimes the speaker prefers to confound the listener. Finally, the investigator may choose not to hear or, if of a different culture, society, or ethnicity from the participator, may have limited capacity to understand nuances.

Personal response is the third fundamental skill. "Talking about oneself to another person is what I call self-disclosure" (Jourard 1964:21). Giving words to our thoughts, feelings, and emotions and telling our hopes, fears, and dreams make us known to and understood by others and, at the same time, enable us to know ourselves better. Self-disclosure has everything to do with relationship – with ourselves and others. With or without conscious intention, we maintain a safe balance between protecting our privacy and revealing our inner selves. For an investigator, of course, giving personal response is an essential qualification for making connections with participators and for collecting rich and reliable data. "Real-self disclosure begets real-self disclosure" (Jourard 1964:64). We move along the continuum of isolation–connection to the place where we feel secure and intact at any moment, depending on many considerations; among the most important are psychological factors. In the following excerpt, a woman explains that she felt such close connection with a participator that it momentarily interfered with her staying in touch with him:

> I was interviewing a homosexual man about his lifestyle. While mentioning the importance of "beauty in his life," Sam added that he often thought that the reason he did not have a lover was because he was not good looking enough to attract a man. I was very touched by this statement for two reasons. First, Sam looked and sounded very sincere, which made me empathize with him. Second, his statement seemed very familiar. It was as if he were speaking my mind. I, too, had often felt that I was not beautiful enough to be loved by a man. Feeling sad and sorry

> for Sam, I tried to protect his feelings and make him feel better. So I looked at him and told him that he was not alone in feeling the way he did; that others felt the same way at different periods in their lives. But what I really had in mind was: "Don't worry, Sam. I'm in the same boat." Needless to say, by sympathizing and identifying with Sam, my mind was transported to the times when *I* had felt like Sam did. Given that my mind couldn't be two places simultaneously, chances are that I missed parts of Sam's comments in the meantime. (Ely 1991:113)

Summary

Qualitative researchers deal with the lived experience of participators; this inevitably includes the non-rational, the emotional and feeling aspects, of the participators' actuality. Investigators must be able to handle all realms of human experience – not to analyze them, not even to understand them, but to be able to listen and truly take in the meaning of the other's lived experience.

The human factors (subjectivity) of the investigator are recognized as the connection or lack of connection with the participator. With this in mind, I have examined the role of the researcher in the light of the mutuality of influence between the knower and the known. The researcher must develop self-awareness, which demands special training. I have specified some personal qualities and skills essential to conscious and responsible research, along with the need for special training directed toward meeting these requirements. Chapter 3 provides educational and theoretical bases for the course that is described later in this book.

Chapter 3

Foundations

A meaning doesn't exist. A meaning is a creative process, a performance in the here and now. This act of creation can be habitual and so quick we cannot trace it, or it can require hours of discussion.

F.S. Perls (1969)[1]

This chapter covers the conceptual background for the methodology and content of the doctoral course "The Role of Self in Qualitative Research."

Confluent education

Confluent education is the approach used to conduct the course. Developed at the University of California at Santa Barbara under the directorship of George Brown, it "describes a philosophy and a process of teaching and learning in which the affective domain and the cognitive domain flow together" (G.I. Brown 1990:7). *Cognitive* refers to intellectual functioning: the activity of the mind in learning about ideas, learning skills, or knowing objects. *Affective* refers to the feeling and emotional aspects of teaching and learning. The approach assumes that an affective component is present in all activities to some degree. Confluent education takes into consideration and *explicitly* uses the emotions, feelings, fantasy, and imagination of the teacher and students in the interaction of teaching and learning to make those activities personally meaningful, thereby expediting and enriching the learning experience.

> At home, first graders watch a Jacques Cousteau television special on turtles. They see the frigate birds eat most of the eggs the turtles lay. The next day in class their teacher has them play the roles of the turtles and the frigate birds. They not only "do" this but also talk about how they feel as they do it. And they talk about similar feelings they've had in other situations. They write stories and learn to read and to spell new words. Can first graders understand tragedy? Can they experience tragedy as part of the condition of nature and life? Can they be stronger for this? And can they learn "reading, writing and arithmetic" as well as they would in a conventional lesson – perhaps better? (G.I. Brown 1990:1)

This experiential approach to learning incorporates basic principles of the Gestalt therapeutic approach believed to be facilitative in this context in three respects: first, intensifying an experience through a here-and-now focus; second, emphasizing awareness; and third, stressing connection with self and contact with the environment – including other people. All three aspects are considered essential to self-scrutiny and attentive interaction with others, and thus necessary for conducting effective research in the field. *Process orientation,* inherent in the Gestalt approach, refers to a person's *becoming* (i.e. developing), over the course of time, through a succession of events that are experienced and responded to. The approach encourages people to open themselves to the actuality of the moment, to the feelings and emotions that accompany experience in a continual flow. To *trust in process* means to have confidence that one's experience gradually becomes something else, without any manipulation to direct it to a predetermined outcome. Each moment blends into the next moment in a natural development. Along with such engagement in life comes a sense of oneself as a person moving, developing, and growing over time, rather than as a static being to be directed and coerced, either by oneself or others. This attitude fits with the evolving nature of an open and dynamic field research situation.

The Gestalt approach

I provide an extensive discussion of the Gestalt approach here because of the uniquely applicable aspects of its theory and practice. This discussion is warranted on two counts: first, the Gestalt approach is consistent with the epistemology and ontology of naturalistic inquiry,[2] (as stated by Lincoln and Guba 1985:37; and Patton 1990:39); second, it provides a theoretical foundation for much of the course for training researchers, described in Chapter 5. The attention on the individual in this chapter relates directly to an emphasis in the course whereby subjectivity, the human factors of the student as potential researcher, becomes the major component of the course.

Gestalt therapy was developed primarily by Frederick and Laura Perls during the years preceding and following the Second World War.[3] Frederick Perls hoped to create a "realistic and effective form of understanding" (Perls 1969), a basic approach short on explanations and long on applicability. His description of effective action as "action directed towards the satisfaction of a dominant need" holds the seeds of much of what follows. I will refer to it in discussing what Clarkson and Mackewn (1993:33) present as his major contributions to theory: holism, field theory, cycle of experience, contact, theory of self, and psychological (boundary) disturbances. There is a close relationship between the ideology and operational principles of the Gestalt approach and those of qualitative methods. They share a holistic outlook. The interaction is participatory; Gestalt's I–Thou connection can be likened to the researcher–participator interaction. The Gestalt therapist develops hypotheses as an ongoing part of a particular therapeutic process; the qualitative researcher develops working hypotheses that apply to a specific case. In Gestalt theory, linear cause–effect thinking is seen as simplistic and ineffectual in understanding and working to alter behavior, as it is in qualitative research in explaining the workings of social actions and interactions. The therapist's and the researcher's values influence their processes, how they choose to see problems, and their choice of clients (in the case of therapists) or participators (in the case of researchers). (In drawing these similarities I have borrowed from "the axioms of the naturalistic paradigm" of Lincoln and Guba 1985:37.) Because of this concordance the training we had developed in the past for Gestalt practitioners was easily modified to train researchers in qualitative methodology.

Holism

Patton lists "holistic perspective" as one of the ten themes of qualitative inquiry (1990:40). Perls objected to dividing a person, or anything else, into "parts." Instead, he acknowledged the interconnectedness of everything in the universe. He described Gestalt therapy as primarily an integrative approach and highlighted the importance of working with the client to bring into dialogue the fragments of the personality as a means toward further integration. For him wholeness was synonymous with wholesomeness, or well-being. He viewed the person, like every other organism, as a being naturally striving to complete what is incomplete in order to achieve integration, satisfaction, and health.

A *gestalt* is a form, or a whole; although the whole may consist of parts, it is more than and different from the sum of its parts. *Gestalt formation* is the organization or configuration of parts into a whole. An "aha" experience is an example of a gestalt formation, of fragments of the *ground* emerging into *figure* as a whole (see "The field," below). Perls used the example of three sticks, which are simply sticks unless one places them together to make a triangle; then they become a whole, more than just three sticks. A school is more than the sum of its building, students, teachers, and other parts. A gestalt may be a tangible thing, such as a triangle, or it may be a situation. A happening such as a meeting of two people, their conversation, and their leave-taking would constitute a completed situation. If there were an interruption in the middle of the conversation, it would be an incomplete gestalt.

It is human nature to have an aversion to an incomplete gestalt. Consider a piece of music cut off before completion, or a person's leaving without saying good-bye. One may physically leave an unfinished task, interaction, or situation of any kind, but it is almost impossible to leave it totally. Emotionally and psychologically it lingers. We long for closure, to finish up what is unfinished before we move on.

Investigators conducting interviews may notice how many people repeatedly mention themes or past experiences that may or may not be directly connected to the inquiries. It is important to appreciate that such a subject is being influenced by a dominant need to manipulate the conversation to deal with her or his unfinished situation, attempting to bring closure to what still demands completion.

Perls's view of health and wholeness includes the ongoing person–environment relationship, whereby the person uses the environment to satisfy needs. Constant interaction occurs not only between a person and other people, whereby emotional needs are satisfied through contact with others, but also organismically between a person and nature, as one takes in air and food and puts out carbon dioxide and other wastes. In all ways human life and the environment engage in a continual exchange.

In contrast to one following the quantitative paradigm, which often investigates isolated variables of phenomena, the qualitative researcher is interested in the whole of whatever phenomenon he or she investigates. Therefore, the qualitative researcher needs to be cognizant of systems and interrelationships, interdependence and mutuality, and the processes through which these are manifested.

The field

"Organization plus environment equals field" (Perls 1973:18). Imagine you are in a large supermarket where the products are jumbled and you have no idea how to find what you need. Or think of a bookstore where the books are shelved not in meaningful categories according to themes and authors, but randomly. You may be aware of your dominant need, but effective action is unlikely. It would be a stroke of luck if you found anything to satisfy you. Stores work for us only if the products are organized and we have experience in recognizing and selecting the items we want.

Researchers approaching their data have reported experiencing similar chaos. How to organize all of this information? In their impatience to make order, they may try to "figure out" what would, with time and immersion, "click into place."

No one organizes our shifting environments for us as we move through life. We do this for ourselves, constantly organizing our fields to satisfy our needs. One's field comprises oneself and one's environment. If we could not organize our environment, we would find it as confusing as the supermarket or the bookstore above. We would become overwhelmed, like the novice observer in the classroom (see p. 15). Fortunately we have a "system of orientation and manipulation" (Perls 1973:19). While writing a grocery list, our consciousness tells us what we require to satisfy basic needs; when we buy new shoes, we can direct our awareness to know which pair is comfortable. Also, we remember what we have learned about satisfying our needs from previous experience so we can avoid people and situations that bring pain and seek out those that bring pleasure. The way we organize our environment shifts according to our dominant need. This is also true for those conducting research in the field.

What is of greatest interest emerges from the *ground*, or background, and comes into *figure*. If when driving along a highway we notice the gas gauge approaching E, we begin paying attention to gas stations; we could then say these gas stations "come into figure." Feeling alone at a social gathering, you spot the one familiar face in the room; the same could be said of that person. In the words of Martin Buber, "No thing is a component of experience or reveals itself except through the reciprocal force of confrontation" (1970:77).

Perls's phrase "the dominance of the most urgent need" means that when we have more than one need pressing for attention, we attend to what is most urgent. The organism "knows" and the person acts. This Perls called "self-regulation," meaning that the organism, which is constantly in a relationship of give and take with the environment, will spontaneously seek out what it requires most from the environment and get rid of whatever it has in excess.

Our emotional needs may not be as obvious to us as our physical needs. We have many ways to anesthetize ourselves, to deny emotions and the bodily sensations that accompany them. Since effective action is action that satisfies a dominant need, we must be in touch with more subtle feelings to have some awareness of emotional needs; otherwise there is little hope that we can make sense of our behavior or find satisfaction. When we need to cry but deny that need over and over again (crying is stupid, it doesn't do any good to cry, crying is for babies, you don't have anything to cry about), we also deprive ourselves of the satisfaction of the organismic release that comes with crying. We are left with an incomplete gestalt.

In reality we have many needs that are never fulfilled. Life is a series of frustrations, interruptions, and unfinished situations. In the dependency of childhood, our basic physical and emotional needs are many and our abilities to get what we need are extremely limited. Our most urgent need in intensely painful or repetitive situations is to anesthetize ourselves, bury the pain and fear, and give up hope. Past incomplete *gestalten* elbow into the present to motivate, blind, and prejudice us with their load of unacknowledged and unexpressed emotions.

The field, in research terms, is the setting where the focus of the investigation is happening, such as a hospital, a school playground, or a family. Researchers, both in their fieldwork and in approaching their collected data, have reported experiencing chaos similar to what is described above. In these situations they need a wide lens and flexible focus. Of primary importance for researchers are the Gestalt concepts of figure/ground; of context (i.e. the field), establishing the meaning of what emerges into figure; and of *allowing* figures to emerge rather than forcing interest or concentration. Discomfort with the surfeit of information and impatience to achieve order out of chaos may drive a desperate researcher to try to figure out patterns, themes, and categories. Time and immersion are necessary for *organizers* to emerge – central themes that the researcher uses to reach an understanding of the data.

Cycle of experience

The satisfying of needs is ongoing in all living things, such as a tree taking in water through its roots, or bees cooling their hive on a hot summer day. The Gestalt *cycle of experience* describes the process whereby people experience and satisfy needs.[4] Imagine a circle representing a period of time with the following sequence of events: experiencing of need, mobilization of energy, contact, satisfaction, withdrawal, and rest until a new need emerges. In life we all go through this sequence of experiences and behaviors, and for us this dynamic process is more complex than that of a tree or a hive of bees, for we have higher thought processes to aid or confuse us. We often interrupt the natural flow of what Perls called "organismic self-regulation."

To achieve satisfaction we have to know what we need. However, a need is not necessarily conscious or clearly conceptualized. We gravitate toward a sunny window without realizing we are cold. A child who needs vitamin C will reach for an orange like an underwater swimmer surfaces for air. Learning to recognize needs is an important aspect of Gestalt therapy and education. Throughout our lives needs arise, and we do what we can to fulfill these needs and effect closure. An unfulfilled need – be it physical, psychological, or emotional – continues to demand our attention. The Gestalt cycle of experience is a frame for conceptualizing one's own and others' behavior. It can help the researcher comprehend field situations that seem vague and incomprehensible. This may be the case, for example, when an interviewer has a particular need (conscious or unconscious) and the participator has an opposing need.

Contact

Contact is the meeting of oneself and what is other than oneself – other people and things in the environment. (To "come in contact with oneself" means to pay attention to what has been outside one's awareness, sensations, thoughts, and so on. When you ask yourself, Why am I doing this?, and give a reply, you come in contact with an intention with which you were previously out of touch.) Contact is necessary for life; we must make contact with air and inhale it, with food and ingest it. It is well known that infants do not thrive without regular physical contact. Families and society often use deprivation of contact with others as punishment: parents send misbehaving children to their rooms, criminals are sent to prison and, in extreme cases, to solitary confinement. Through contact, events take place. Without contact, there is no awareness, no energy, no dialogue, no process.

Perls emphasized the importance of an "appreciation of differences" as a requirement for contact. Between persons this means that when I am in contact with you, I have a sense of myself and an awareness of you as the *other.* I am aware of, and own, my physical sensations, thoughts, and emotions. In Gestalt terms I take *personal responsibility* and recognize that I am I and you are you; we are separate entities with boundaries. In this sense *boundary* is not a real thing or place, but a metaphor for the division between what is oneself and what is not oneself, including the *physical* dimension – my hand differs from your hand and we can feel the difference; the *emotional* – I experience my tenderness and you experience yours; the *mental* – I formulate my ideas and you formulate yours; and the *spiritual* – I experience my faith and universal connection and you experience yours.

Contact, in the Gestalt view, results in change. When one enters into each moment with an attitude of potential acceptance of and total involvement in whatever may occur, there can be complete contact. Then, through the totality and integration of experience, change – unpredictable change – is possible. If one approaches an experience wanting some kind of transformation to happen, making comparisons with what was or what should be, this mental activity will interfere with total contact and experience. Wanting a specific outcome usually implies trying to manipulate a situation rather than trusting the process of contact and involvement. Researchers who "know" in advance what their subjects will say or what their investigations will show may find what they are looking for. However, they miss much relevant, perhaps contradictory, data that would yield new and important information they had not considered.

The ability to make aware contact is crucial for the researcher in relation to participators and environments in a study. Heshusius refers to "a participatory mode of consciousness on the part of the researcher" (1994:16). (See also p. 29.) The experience of "participatory consciousness" involves a knowing of the other, which is difficult to describe because it is an unusual happening. One temporarily relinquishes "egocentric thoughts, feelings, and needs" (Heshusius 1994:18). In this state one temporarily achieves what Buber called the "I–Thou" connection with another, and the potential for unpredictable change, mentioned above, exists.

Theory of self

As described earlier, attention to subjectivity, to the self of the researcher, can lead to more complete and accurate data gathering. It can also lead to the development of the person of the researcher, and thus to "sharpening the instrument." Consequently, a fuller explication of self-in-process and what might interfere with its healthy development would suit our purpose here.

Perls objected to the widespread practice of "making a thing out of a process" (1969). In his view, self is a process that is manifest when one engages with ideas, nature, and other people – in a word, with all that is separate from oneself. The word *self* denotes not only a person's physical body, but primarily one's subjectivity, points of view, and suppositions. According to Berg and Smith, "It has always been hard for social investigators to know what to do with their selves ... we have to make a choice about where we will position ourselves" (1988:9). They refer to the degree of personal engagement, the influence on participators that researchers inevitably exert, intentionally or unintentionally.

In the framework of Gestalt therapy, one makes contact with what is not oneself. Inherently, then, a conception of boundary exists between what is and what is not the self. In Gestalt terms I take personal responsibility and recognize that I am I and you are other; we are separate entities; we do not merge. (There are times of apparent, temporary merging: Buber describes an "I–Thou moment"; people report "becoming one with the universe"; Heshusius writes of "participatory consciousness"; lovers speak of "becoming one.") Heshusius notes that "the capacity for autonomy is not relinquished," although you forget yourself (1994:18). This becomes a critical point when we speak of personal responsibility, of I as an agent (*I* choose to sit here, *I* make myself feel guilty, *I* wrote that letter, *I* am angry, *I* don't like what is happening), as opposed to *boundary disturbances*, where the agency of actions and ideas is confused. (See below.)

The self-in-process organizes the supermarket of life, especially at times of "particularly intense needs and activity, possible great stress or heightened emotions" (Korb et al. 1989:29). Some years ago when an airplane crashed into a partially frozen river, a passerby spontaneously jumped in and pulled out one of the victims. One could cite many such examples of someone acting without thought or consideration, choosing an action in response to their own or someone else's danger. According to Korb et al., "In such a time, persons experience contact with surroundings in sharply defined terms and experience the evaluation of what is appropriate or true with clarity and immediacy" (1989:29).

For Perls, actualizing oneself was not a program of self-improvement. He referred to this common interpretation as "*self-concept* actualization." He understood actualizing oneself to be a natural process of a person's development over time, of becoming, of maturing: "self-actualization means the wheat germ will actualize itself as a wheat plant and never as a rye plant" (1969). Would that a human being could grow and mature as simply as a wheat plant! For many of us, it is enough to get back on a track where we can connect again with ourselves, experience ourselves in contact with others, and dare to have moments of participatory consciousness.

> Ultimately what's most lasting are those intimate moments that are so fleeting. Those magical, mystical moments where man leaves himself and becomes one with another person. There is no witness except the person you shared it with – a woman, a child, another man. I think if you stacked all those tender moments head to head, if you get out of life with twenty-seven minutes like that, you're probably way ahead. (Jonas Salk)

Although not as intense, connecting with respondents in a research project can bring not only energy and satisfaction to all concerned in the moment, but also the possibility of a worthwhile research study.

Boundary disturbances

The following discussion deals with projection, confluence, retroflection, introjection, and deflection – psychological strategies frequently used to interrupt contact with oneself and others. These dynamics are not always negative; for example, projection allows one to project oneself into another's situation and imagine it as one's own, possibly resulting in empathy. The following explains negative aspects of each disturbance separately, although they often seem to be combined.

Projection

Assume we are talking together. I listen to you, even when it may be unpleasant for me. I become angry but do not admit this to myself; rather, I begin to think *you* are angry with *me*. I have unconsciously disowned my anger; in my experience it no longer resides in me. I have placed my emotion outside my boundary into you. Through my lack of self-awareness, the boundary between us is now blurred, for what is mine I experience as yours. As a result of this dynamic, I am in contact with neither myself nor you. This is an example of projection. Without awareness researchers sometimes project their own emotions, feelings, thoughts, and values onto the people or groups they are studying, thus losing their appreciation of differences. To some degree the participator, or interviewee, is no longer the other. The data collected will suffer in quality; they may reflect more of the researcher's than the participator's story.

Confluence

Listening as you relate a recent experience, I may respond with deep sympathy and empathy, yet I recognize that I am still separate. You have your feelings and I have mine in response to you. Such a clear separation between self and other is essential for contact and for collecting valid data. However, if I feel so open to your emotions that I lose all feeling of separation, in some way I merge into your emotions. Then I am in confluence with you and not in contact, for contact requires a sense of I-ness and other-ness and I have lost that. You are no longer the other. Researchers have been accused of "going native" when they take on the dress, values, and habits of the people they are studying. These may be signs of confluence, of the researcher's having lost sense of him- or herself as separate from the subjects. Under such circumstances it becomes difficult if not impossible to maintain the openness necessary to collect accurate data and interpret them meaningfully.

Retroflection

This occurs when we do to ourselves what we would like to do to others. Perhaps in an effort to be nice to an interviewee, the researcher turns any hostility he or she may feel back toward her- or himself: I am a bad person, I shouldn't feel this way, I shouldn't be doing research. Such confusion makes openness and contact unlikely.

Introjection

We do this when we "swallow whole" attitudes, ideas, and values, especially from parents, school, and church during our youth. As adults those who have never examined the commandments that rule their lives commonly utter empty proverbs and aphorisms, not only to themselves, but very often to others as well. New situations threaten the introjector because he or she has become incapable of authentic response. Such a stance would never work for the researcher because it would interfere with the availability of the researcher as his or her own instrument. Equally important, the task of "chewing up" information, mulling over and digesting data for meaningful evaluation, would be impossible for the introjector, who impatiently gulps and devours.

Deflection

This boundary disturbance is the avoidance of whatever comes toward one, whether words, looks, or gestures, as if one carries a shield; nothing gets through. No contact with others goes hand in hand with no contact with oneself. Deflection is a form of global protection from strong experience – no laughing, no crying. A researcher needs to experience what is going on with an unguarded, permeable, and sensitive attitude. An investigator who avoids direct contact is like a chef who has no taste for food.

In each type of boundary disturbance, people are unaware of their actions, reactions, impulses, or emotions. Through the psychological dynamics of projection, confluence, retroflection, introjection, or deflection – or a combination – they unconsciously blur the boundary between self and other, making contact improbable.

Although I have discussed the types separately and they may appear simple, real interactions cannot be conveniently broken down into distinct, analyzable bits. The mutual influence of different disturbances is hard to pin down; it is more than the words spoken, the glances cast, and the tones of voice. Even if one were to film an interaction and analyze each moment of behavior, the complex web would not give up its mysteries. A human relationship is a nidus of complexities, a wilderness of emotions, a mysterious occurrence that defies precise, comprehensive analysis. Recognizing this reality may help a researcher tolerate formidable situations.

The attitude of Gestalt

Claudio Naranjo emphasizes the *attitude* underlying the work of Perls, with whom he had close association in the 1960s: "These three – an appreciation of *actuality*, of *awareness*, and of *responsibility* – constitute the core attitude of Gestalt Therapy" (1993:6). Naranjo makes clear that this attitude is not based merely on some ideology but is anchored in actual experience:

> The *evidence* of actuality (i.e. the apprehension of the fact that we are living here and now and we are one with our concrete actions): the evidence of responsibility (the fact that we do what we do and that we are no different from what we are), and the evidence of awareness (that we at some level know what we are doing and experiencing, no matter how much we lie to ourselves pretending that we do not). (p. 18)

The appreciation of actuality refers to Gestalt therapy's present-centeredness, for it is only in the present moment that contact with what is – with oneself, with others present, and with the environment – is possible.[5] It also means that the therapist accepts clients as they are rather than being oriented toward changing them. This may seem a paradoxical stance for a therapist; however, the therapist trusts that the very process of living through experiences together will result in growth and change.

The Gestalt therapist, through her or his behavior, exemplifies a concern for the "now," rather than the past or the future; for here, rather than elsewhere; for experiencing what is, rather than imagining; and for taking responsibility for one's actions, feelings, and thoughts (Naranjo 1993:13). This may be seen as an implicit agenda, since clients learn from the example set by the therapist. The same may be said of the research interviewer and those being interviewed. Students who have taken our course have commented on the usefulness of present-centered awareness during interviews for their research studies.

Affinity with qualitative methodology

There is considerable similarity between the Gestalt approach and qualitative research methodology. (Similarities may not be restricted to qualitative approaches, since similarities and overlaps exist between quantitative and qualitative methodologies; however, I will limit this discussion to the Gestalt–qualitative comparison.) Although the goals of the Gestalt process are explicitly therapeutic and thereby differ from those of qualitative research, there is remarkable equivalence in the conduct of the practitioners and the procedures of Gestalt and qualitative inquiry. Both the Gestalt therapist and the participant observer in a qualitative study make self-awareness and awareness of client and context top priorities; each attends closely to become explicitly aware. Both employ a non-judgmental, exploratory attitude toward the client and an active and participatory style. Both follow closely the process of the client, basing interventions on the client's behavior; each response from the client prompts a response from the practitioner, who forms and re-forms hypotheses as he or she gathers information and the interaction continues. The Gestalt practitioner and the researcher share an exploratory, change-oriented approach to procedure and goals. As in Gestalt work, "qualitative research emphasizes social context, multiple perspectives, complexity, individual differences, circular causality, recursion and holism" (Moon et al. 1990:364).

The "Ten Themes of Qualitative Inquiry"[6] listed by Patton (1990:40) are further evidence of parallels with the practice of Gestalt therapy:

1. naturalistic inquiry
2. inductive analysis
3. holistic perspective
4. qualitative data

5. personal contact and insight
6. dynamic systems
7. unique case orientation
8. context sensitivity
9. empathic neutrality
10. design flexibility

1. Naturalistic inquiry

Qualitative inquiry occurs in neither a laboratory nor a contrived situation in which a particular event is controlled. The researcher wants to understand the participators individually or in a group in a particular setting, without "predetermined goals or outcomes but focuses on the actual operations and impacts of a process, program, or intervention over a period of time" (Patton 1990:42). A Gestalt interaction usually occurs in a particular meeting place, like an office or a seminar room, using "methods of present awareness and phenomenological investigation" (Clarkson and Mackewn 1993:84). The therapist engages and thus witnesses the client in relationship – human interaction in which the client's behavior patterns become obvious. The therapist seeks to make interventions that will address psychological disturbances revealed during the interaction. She or he perceives and encourages the client's ongoing process, open to whatever emerges with no predetermined outcome.

2. Inductive analysis

Patton defined this as a process of exploration and discovery without "imposing preexisting expectations" (1990:44). In Gestalt practice meaningful behaviors, patterns, and themes emerge during the interaction between therapist and client and lead to the framing of hypotheses that are then tested and made more explicit and conscious. Qualitative research employs a similar open-ended approach. Through the process of observing or interviewing or examining documents, each piece of information and each encounter directs the researcher's next step. The path is not laid out beforehand with the aid of a theory to prove or disprove. Themes, categories, and patterns emerge in a recursive manner as data are gathered and confronted again and again.

3. Holistic perspective

"The *whole* phenomenon under study is understood as a complex system that is more than the sum of its parts; focus on complex interdependence not meaningfully reduced to a few discrete variables and linear, cause–effect relationships" (Patton 1990:44). This description coincides with the Gestalt perspective. Perls was averse to analyzing a piece of behavior and losing sight of the whole individual, or seeing a limited context and ignoring the larger field: "The individual is inevitably, at every moment, a part of some field. His behavior is a function of the total field, which includes both him and his environment" (Perls 1973:15–16). In the Gestalt view, ground gives meaning to figure: "a meaning is created by relating a figure, the foreground, to the background against which the figure appears" (Perls 1969).

4. Qualitative data

In research, these are data rich in descriptions and quotations rather than reasons, explanations, interpretation, or analysis. In Gestalt, involvement of feelings, emotions, and physical expression provide vivid description of experience and reveal the extent of engagement; explanations, rationalizations, and justifications are irrelevant. Perls wrote, "I rely upon the patient's detailed descriptions of his experiences and my observations, and try to use as little construction and guesswork – for instance interpretation – as possible" (1979:13).

5. Personal contact and insight

Like qualitative researchers, Gestalt practitioners use themselves as instruments in gathering information in present-centered interaction and in providing responses to what is going on. The Gestalt therapist, whether active or quiet, becomes personally involved with and responsive to the client. Contact, of supreme importance in the Gestalt process, is emphasized between therapist and client, client and her- or himself, and client and context. The researcher, meanwhile, has direct contact with participators in their customary settings and routines. By being in the field, meeting people, observing, and interviewing, the researcher encounters a context that provides meaning for the observed behavior.

6. Dynamic systems

In qualitative research this term refers to the assumption that change is inherent and ongoing; such research always focuses on the moment-by-moment unfolding of experience. This corresponds exactly to the emphasis on process in Gestalt. Trusting one's experience in constantly changing reality is a risk that clients are encouraged to take in the course of their work.

7. Unique case orientation

This allows in-depth investigation into individual cases. The same orientation is intrinsic in the Gestalt approach; the therapist deals with present phenomena of the particular person in the moment, neither fitting the client into a diagnosis nor making comparisons or generalizations. This significant feature of qualitative methodology leads to discovery of what happens in a particular setting, uncovers new knowledge, and builds theory.

8. Context sensitivity

What is discovered in one context in qualitative research is not assumed to be true in another context; the investigator does not generalize "across time and space" (Patton 1990:40). The Gestalt approach makes clear that the context gives meaning to what is in figure; a change in context changes meaning and appropriateness of behaviors. Through Gestalt experiments, with individuals and in groups, one can expect an increase in clients' understanding of others' contexts through role-playing and observations.

9. Empathic neutrality

Patton wrote: "Complete objectivity is impossible; pure subjectivity undermines credibility; the researcher's passion is understanding the world in all its complexity – not proving something, not advocating, not advancing personal agendas, but understanding; the researcher includes personal experience and empathic insight as part of the relevant data, while taking a neutral nonjudgmental stance toward whatever content may emerge" (1990:41). The Gestalt therapist takes an impartial, non-judgmental stance toward whatever content may emerge from the client. The therapist may be passionately involved yet still detached from both the client and the outcome of the therapeutic process.

10. Design flexibility

In both qualitative inquiry and the Gestalt process, this denotes an openness to adapting and adjusting direction and manner in response to the client's words and/or behavior. The therapist makes interventions that rest on hypotheses based on the words and actions of the client. The lack of a specific outcome – in qualitative research as in Gestalt therapy – does away with a rigid program or manipulation. The interchanges are spontaneous, but the procedures are not random; they are built on the foundations of Gestalt theory and principles.

This striking correspondence between Gestalt practice and qualitative methodology allows us to take advantage of the basic training scheme we have developed over many years for the Gestalt approach, most of which can be readily and directly applied for use in training researchers in qualitative inquiry.

However, this does not suggest that one must become a Gestalt therapist to carry out qualitative research. Rather, we borrow those aspects that are directly transferable, including – but not limited to – basic awareness training, holistic perspective, process orientation, and the paradoxical nature of change.

Paradox

"There is something in the nature of paradox," wrote Watzlawick et al., "that is of immediate pragmatic and even existential import for all of us; paradox can not only invade interaction and affect our sanity but also it challenges our belief in the consistency, and therefore the ultimate soundness, of our universe" (1967:187).[7] The concept of paradox is implicit in the theory and practice of Gestalt therapy. To be human is to experience contradictory feelings, emotions, attitudes, and beliefs – simultaneously – at different levels of our being. Inherent in the Gestalt approach is an appreciation of the paradoxical nature of human beings, the impasse that results from holding or denying contradictions, and the limitations of addressing only the cognitive realm when dealing with issues of human communication, growth and change. Understanding paradox helps illuminate many puzzling human phenomena that seem to defy common sense or logic.

Logically one would expect New Year's resolutions and self-improvement plans to work. That is, people who are motivated to change their behavior would be successful in their attempts. A typical example: "I should go to bed earlier so I can get up earlier and get more done." Perls understood that most schemes to bring about behavioral or attitudinal change awaken an opposite and equal force in a person, resulting in a struggle, but no change. On the heels of "I should go to bed earlier ..." follows a contradictory statement, "I hate to get up early," and the resulting circularity of "I should"–"I won't." This leads inevitably to an ongoing and familiar contest and, contrary to what one might expect, no real change. A paradoxical situation; an impasse.

Another example of a paradoxical frame in Gestalt is the

therapist's attitude and role. The therapist accepts the client as he or she is rather than being oriented toward changing the client. This attitude rests on the belief that change comes when clients are just as they are, do what they do, and experience what is happening in their lives, always with awareness, *never with expectation of change*. Growth and change occur through engaging in one's process and experiencing what is; they are not forced to happen through design.

The following illustrates working with a split in a therapeutic context – the self as resentful of father and the self as forgiving:

CLIENT: My father's been dead for years and I'm still talking to him in my head every day.
THERAPIST: Please, imagine your father, not in your head but in this chair that I am placing opposite you. Tell *him* what you just told me.
CLIENT AS SON: Dad, I talk to you every day. There's so much we never talked about. I have resented so much that you never talked to me about Mom. [*With tears in his voice.*] I know it was hard for you when she died, but I was only a kid of 17 and I needed to know what was going on. I needed someone to talk to. You were the only one and you were so shut down. You really left me on my own.
THERAPIST: Change chairs now, please, and be your father.
CLIENT AS FATHER: [*Sits in silence for a moment.*] I don't know what to say. I guess I was so numb I didn't notice anyone else. I was devastated. [*Begins to cry.*] I don't think I ever got over your mother's death. Oh God, it was awful. [*Cries harder, then crying subsides.*] I do see you now, though. What a tough time that must have been for you. I am sorry you had to go through it.
CLIENT AS SON: [*Cries quietly.*] I'm glad to talk to you out loud, finally. And to hear your side of it. I have resented you so much. You know, I even blamed you for Mom's death. Now all that seems to disappear. I just wish we could have talked years ago. If you were really here, I would like to embrace you. [*Takes a pillow and hugs it. Cries deeply and looks up at the therapist.*] You know, I really love my father.

For some it is a revelation to discover they have within them strong ambivalence, sometimes even warring sides. In Gestalt terms the client begins to experience, identify with, and own his or her oppositional material, no longer projecting it onto someone else (or some institution) who becomes the adversary. Paradoxically, what is hidden has great power; what is revealed and owned loses its hold on us. That which originates at an earlier life stage as a psychological safety maneuver (disowning thoughts, desires, emotions) becomes dangerous for oneself and others as time passes. Reowning the disowned material is seen in a therapeutic context as a step toward integration as well as a safety measure. Disowned emotions and characteristics can become dangerous when activated under certain conditions. Perhaps this is what our legal system calls "temporary insanity."

A paradoxical perspective dispenses with linear thinking and problem-solving approaches. It allows what have been defined as splits between conflicting elements to be reframed from a metaposition as linkages. With conflicting elements recast in this way, a new level of meaning can come out of the creative tension between them, as demonstrated in the son–father example above. The paradoxical view recognizes the coexistence of multiple meanings of an event or circumstance. In striving to succeed in a given situation, success may come from an unpremeditated, spontaneous act after all deliberate efforts have failed. Imagine a doctoral student whose study involves interviewing school principals. He finds he experiences the same anxiety he had as a child when called into the principal's office. He forces himself to continue, knowing his reactions makes no sense, but his interviews fail. He soon decides to quit. "Nothing is worth going through this," he tells himself. "I'll never get my degree. I'll never be a researcher." Yet he decides to go to one more interview that he had already arranged. He approaches the office door with an "I don't care what happens" attitude. He finds he is relaxed, because he really does not care what happens. The interview goes very well. By giving up, he succeeds.

Where flexibility and openness are important features, as they are for the researcher, incorporating the paradoxical perspective can be instrumental in accepting contradictory data, being open to conflicting hypotheses, reframing and metaprocessing, and, finally, in maintaining a professional disidentification with the field and its participants while at the same time being present, engaged, and responsive to what is going on.

Gestalt and education

The theory and basic attitudes of the Gestalt approach have, as mentioned early in this chapter, implications for educational practice in two respects: first, a focus on the affective domain – emotions, values, and attitudes – together with cognitive learning; and second, a present-centered focus. Subsumed under these two focuses is a concern for personal growth of students and the primacy of relating (existential connection) to the self, to others, and to the subject matter. A confluent teacher fosters awareness, personal responsibility, and focus on actuality. She or he also structures experiences relative to these fundamental features of the Gestalt approach. In our course we have found these positions to be directly relevant to developing personal growth and the abilities to relate to others and to evaluate and analyze data.

Affective domain

Affective elements are inherent in all learning situations and must not be ignored.[8] When a student requires particular practices and skills, then learning by doing is essential, and any emotions, values, and attitudes that accompany the doing should be acknowledged. The intention is for feelings, fantasy, and imagination to facilitate, clarify, and make all learning personally meaningful. Even abstract subject matter, such as mathematics or physics, which may seem so dry to an outsider, can evoke the learner's interest in, even passion for, the challenging problems they present.

Figure, ground, and meaning

Perls realized the applicability of Gestalt theory to education. His definition of learning was "to discover that something is possible." The word *discover* is important because it signifies a happening in the present, based on awareness. Making meaning hinges on the relationship of figure and ground – of what is in focus in relationship to the context. We make errors when we lose sight of the present context, and make meaning on the basis of previous experience or context.

Consider a story I heard in Holland many years ago. It was spring, near the end of the Second World War, and people were indulging in a Dutch favorite: raw herring sold on the street, eaten with the fingers and with the head thrown back. An American journalist observed this and wrote back to the publication for which he worked: "You can't imagine the starvation these people have known. They can't wait to take the fish home and cook them, but eat them raw in the streets!" That the Dutch had suffered extreme food shortages during the war was true; however, that fact had nothing to do with the local custom of eating raw herring on the street.

Needing to explain this strange phenomenon, to find meaning, the journalist fell into the trap, especially dangerous for researchers, of being self-referential. For him the only condition under which *he* would have eaten raw fish was starvation; reasoning that these people were like him, he wrongly concluded that starvation was the reason behind their seemingly strange behavior. He depended on what he "knew" from a different context and construed the same meaning regardless of the given context. The story illustrates that seeing what is in figure is not enough. Quick interpretations, assumptions, and prejudices keep us from learning that something new is possible in any particular context.

The unconscious is a powerful force in determining how and what we see, and how we organize our experiences and reactions. Situations that remain incomplete because we have been unable to bring them to closure may result in *fixed gestalten*, as in the case of the doctoral student interviewing school principals. Another example is a father who sees his son as a troublemaker. No matter what the child does, the father comes up with the same conclusion: the boy is a troublemaker. Why the father needs to see his son this way and how he came to be like this are matters of speculation. We might hypothesize that the father's father did the same to him, and/or that the parents need a scapegoat to take attention away from their unhappy marriage. On the basis of Gestalt principles, one could say the father, needing to release tension derived from unaware needs, misdirects his anger and aggression and never achieves satisfaction. At any rate, the fixed gestalt blocks his ongoing process and precludes learning, or the discovery of anything new. He simply repeats the same behavior.

Obviously, how one organizes an experience affects one's behavior. The journalist writes his story and never finds out the truth: that in spring the small migrating herring are caught in the vicinity and are considered a delicacy. Nor does he discover for himself how delicious they are. The father, with some willingness, help, and much courage, might become aware that his anger has nothing to do with his son but is confused with feelings about himself. With more clarity he could see his son as a normal child acting his age. The father might build a more growth-producing relationship with both himself and his son. This is possible, however, only if he lets go of his fixed gestalt by allowing a new point of view that would bring a new experience. Healthy functioning consists of dynamic figure formation, of construction and destruction of gestalten based on response to present awareness.

Openness to new meaning is essential in any learning experience; it is especially relevant to the researcher who might be tempted to construe meaning to suit her or his needs, desires, and hopes. Given that learning in Gestalt terms involves discovery, the work of the teacher is to help students engage in the dynamic process that leads to, and includes, discovery. This often involves "unlearning" what people already "know" in order to free them to discover what emerges in the present moment in themselves and their environment, and what is transpiring between the two. Unlearning means letting go of old gestalten – sometimes reawakening buried emotions – and coming fresh to a situation; then new figures and new meanings emerge.

A confluent lesson

An example of the integration of affective and cognitive aspects of learning is the film *Lessons from the Dead*, made at the Medical School, Erasmus University, Rotterdam, Holland. The film shows a class of medical students who are to observe their first autopsy – always a highly charged experience. The professor, Marco de Vries, MD, a confluent teacher, not only lectures on the subject, but also involves the students personally. He had heard previous students tell how all their attention during the autopsy had been diverted to "keeping control of themselves." When they left the laboratory, they remembered little of the autopsy. He wanted to minimize interference to learning during this important demonstration and to increase in any way possible the students' ability to connect with the lesson and to benefit, intellectually and emotionally, from the opportunity. Therefore, he structured the class to meet not only the intellectual needs of the students (i.e. to learn anatomy and pathology from firsthand observation), but also their emotional needs.

Before the students observe the actual autopsy, Professor de Vries allows time for them to explore their feelings about life and death, autopsies, their expectations of themselves as healers, and any emotions they may experience in anticipation of the day's activities. He has the students create role-plays in which a doctor requests permission of a bereaved family member to conduct an autopsy. Each student has an opportunity to play both roles – the doctor and the family member.

The film continues in the laboratory, showing an actual autopsy with accompanying commentary by the pathologist in charge. At the completion of the laboratory segment, the students return to the classroom, where they get an opportunity to talk about their reactions. de Vries also participates, disclosing his experience when he asked the wife of the deceased man for permission not only to perform the autopsy, but also to use it as a demonstration in class and to film it. Her husband would have been pleased, she told him without hesitation, explaining that he had wanted to donate his organs to science but that, because of his cancer, this had not been possible. Now his wish to contribute to others' learning would be granted. She also described what a disciplined, hard-working man he was and how he loved to tend his garden; he thought people should be responsible and contribute to society.

Watching the film one can see the rapt attention of the students and their emotional responses as de Vries relates these facts. As in all messages there is the content level (i.e. the facts) and the metalevel (i.e. the message of the message). The metamessage deals with relationships: the body on the autopsy table, which appeared as merely an object in the context of the laboratory, juxtaposed with the mental picture of the live, thinking, feeling human being; the man as husband to his wife and in relation to other people; and the man working in his garden.

Helping students connect personally with subject matter creates a lively and meaningful learning situation. The students' learning, in addition to being enriched, becomes intensified far beyond mere subject matter. Use of fantasy and exchanges of ideas with other students and the teacher broaden and deepen the acquisition of knowledge. One comes to appreciate an expanded range of intellectual and emotional reactions, in oneself and others. In the case of the medical students, one can presume increased openness to their own emotions, those of their future patients, and those of the families they might one day have to approach with the request to perform an autopsy.

It is not difficult to transfer the nature of this situation to similar emotionally loaded situations encountered by researchers; by becoming more open to their emotional responses, they can use their selves as sources of data. Thus, our course (Chapter 5) pays much attention to this issue.

For effective learning, both teacher and student must develop awareness of

- needs (i.e. clear figure formation);
- presence and engagement;
- energy mobilization;
- where to find or how to create in the environment what is needed for satisfying contact (e.g. a book, a person, a place, or maybe some affection); and
- when successful completion is not achieved and how to take another direction for better contact and satisfaction.

For researchers as well as teachers, the Gestalt approach keeps in focus motivation, inherent satisfaction, and personal responsibility in all interaction.

Summary

This chapter has introduced confluent education, with its rich inheritance from the Gestalt approach. It brings to the classroom a balance of cognitive and affective learning. I have surveyed the basic principles of the Gestalt approach, including holism; the field (or context), which lends meaning to what comes into figure; and contact. By juxtaposing Patton's "Ten Themes of Qualitative Inquiry" with the guiding themes of Gestalt practice, I have shown how they parallel each other. Although the two approaches are used in different contexts with different aims, the similarity between them is striking. As a way to conceptualize and deal with human behavior, contradictions, conflicts, and impasse, the paradoxical perspective has been considered. It fills the gap left when a logical, linear, cause–effect approach is rejected as simplistic and ineffective. In view of the significance to the qualitative researcher of contact with self and with others, I have explicated the psychological disturbances of projection, confluence, retroflection, introjection, and deflection.

Chapter 4

Metaprocess

[The] ability to meta-communicate appropriately is not only the conditio sine qua non *of successful communication, but it is intimately linked with the enormous problem of awareness of self and others.*

P. Watzlawick (Watzlawick et al. 1967:53)

Relationship

Metaprocess is a particular kind of human communication. Therefore, before rushing into the main subject of this chapter, I must comment on *relationship,* which is the medium in which communication is rooted. (This statement can also be turned around: relationship is rooted in communication. Each is a function of the other.) Relationship claims a central position in social science research and demands special skills and qualities of researchers. The key to the integrity of research, regardless of methodology, is the researcher. Where the researcher is also the primary instrument, as in qualitative methodology, validity and dependability rest on the integrity of the researcher. Given the complexity of human interaction and the desirability of maximizing clarity when collecting data, it is essential to consider what is involved in the human encounters from which the data emerge.

Investigators must be able to pay attention to themselves and others, to listen with interest and empathy, and to respond in the moment. While interviewing, the researcher must not box him-

or herself into playing a role but, instead, be available for human-to-human connection. Interaction is the meat and potatoes of interviewing, of people talking to people about their experiences and their reactions to those experiences, their emotions, opinions, attitudes, and beliefs. Communication entails much more than simply sending out and receiving verbal information. It is how people relate, forging the pathway to connection with each other. Relating involves major issues of existence, such as trust, disclosure, isolation, and empathy. Communication in which people invest themselves fully is without pretense, without hidden agendas; it is lively, constructive, and sometimes humorous: a contactful and enriching experience. The participants are *present,* and their responsiveness reveals their presence.

The ability to respond in a situation rests on one's being in contact with oneself. But how else might we be if not in contact with ourselves? Think back to the researcher who returned to his island homeland in Chapter 2 (p. 24). Who was he when he "interviewed" the islanders? He was *playing the role* of researcher. His behavior was an act; he was oblivious to his goals, desires, prejudices, and emotions. He was not in contact with himself or others. A role is a body mask to keep us unaware. We are also oblivious when we slip into profound self-absorption, or behave in response to an emotional fragment, as when a parent becomes enraged with a child and loses all control.

Awareness

Since awareness is the essential ingredient for relating, it is important to take a short bypass before we return to relationship. Through awareness we know what is going on within and around us. We access the actuality of our experience of the outer world through our senses: sight, hearing, taste, touch, and smell. We access internal experiences through the inner proprioceptive senses: posture, movement, pressure, and muscular tension (de Vries 1993:25). Awareness is not cognitive; it is not figuring out, justifying, rationalizing, nor explaining our own and others' behavior. Nor is it dwelling on the past, anticipating the future, or just plain daydreaming. Awareness is attending to what is happening moment by moment. It may be deliberate, or *directed awareness*: for example, you may notice now how you are breathing, how the light falls on the page as you read, or whether you have tension in your shoulders; or it may be *undirected awareness*, as when the smell of something burning on the stove in the next room cuts through into your awareness while you are reading, or an angry voice at the next table in the restaurant drowns out the people you are sitting with, or the facial expression of the person you are talking to suddenly emerges into focus, or cold air smarts your cheeks as you walk down the street.[1]

Self-awareness is reflexive; that is, the subject and object are the same. *You* become aware of *yourself*. It involves neither cognitive nor verbal activity – it does not entail thinking about or explaining yourself: "I am feeling so upset. I don't know why. Maybe it's because he reminded me of my father. But he seemed nice, all the same...." Such mental activity separates you from your experience by sidetracking you into your thoughts. Contrast the following approach with the one quoted above: You become aware of feeling upset and take time for direct contact with the feelings, emotions, and physical sensations that are surfacing from moment to moment, coming to a knowing through focused perception of yourself through your senses.

Ideally we would always be aware of ourselves and others, staying in touch, available for contact and communication. Experience teaches us this is not the case. When people we love or even strangers say things that touch tender spots, we feel attacked in some way and "lose our cool," losing with it the awareness of ourselves and others in those moments. Swept out of the here and now, we fall back on habits that, although familiar, we know do not serve us well. Some people verbally lash out with automatic responses. Some cry. Some beat a hasty retreat, or remain in the room but retreat by withdrawing into themselves. Some apologize, plead for love and forgiveness, and make promises. Reactions that seem senseless, exaggerated, or inappropriate arise not from today's events, but from "bad" habits, ways we learned to handle physical or emotional attacks at an early age – ways based on outmoded fears. Such reactions give emotions a negative tinge, for our reactions are triggered by our emotions. Yet emotions are natural responses to events, not something of which to rid ourselves. "Without emotions we are dead, bored, uninvolved machines" (Perls 1969). And without connection with oneself, there is no self to make connection with another.

This leads us back to relationship. We establish relationship through our exchange of words, expression of attitude, behavior, and tone of voice. A conversation can be so engrossing that we forget about time and place. Yet occasionally one makes a connection with oneself and another at a level other than the topic of conversation: "It's good to talk to you about these ideas." "Yes, I realize how excited I am when we get into these topics." During such interludes the experiences of "I" and "you" come into focus and we make ourselves known to one another through *metacommunication*, that is, communication about our communicating. Our experience *in the moment* becomes the content of the conversation.

In many interactive situations establishing or defining relationship is the primary issue; the ostensible topic of conversation is merely the vehicle to examine or clarify what is happening between the people. For example, a teacher goes to the office of her newly appointed principal with a question about a letter that had been sent to all teachers in the school. Her previous meeting with him had ended rather abruptly, and she had felt brushed off. She might have asked another teacher to clarify the letter, but instead she goes directly to the top man. More important to her than getting the information she needs is the establishing of relationship. She wants to impress upon him who she is; she wants him to take notice of her. By going to him directly, she hopes to convey that she is not afraid to demand his time and attention. The principal has no idea of her motivations. He may or may not guess what she really wants. Regardless of how he responds, however, the face-to-face contact helps to create and define their present relationship. This occurs often in families and in work and social situations. We want to create and define our relationships, to know whether we can make an impact, who is in charge, and whether we are wanted and valued. In such an exchange we have a desired outcome, an agenda, sometimes hidden even from ourselves. Attitudes – positive or negative – that we have suppressed betray us in a slip of the tongue or a certain tone of voice, and we express the attitudes in spite of ourselves.

Sometimes we imagine or sense danger in an interaction, believing the other's existence in some way threatens our existence. Then we might lose awareness of what we are doing. Although not really a matter of life or death, the situation may be charged with such emotional intensity that we feel we are on the brink of a precipice. As if in a time warp, powerful reactions from our earlier years are triggered and strike in a flash. Thus, two people in an interaction may know little or nothing of what they are saying, their physical sensations or needs, what they express by tone of voice, or how they participate in creating what may become a highly disagreeable episode. Both may wonder, How did we get here? What happened? In such interactions we lose connection with ourselves and the other in those moments; relationship issues become survival issues; our emotions block awareness, and the outcome is unpleasant or even disastrous.

Self-awareness and awareness of others lead the way to presence, responsiveness, direct communication, and contact – provided one is able to put words to one's experience. However, *awareness of others* depends in large part on using our eyes and ears as well. We listen to the content of the words spoken; we hear the tone of voice and mannerisms of speech; we see facial expressions, gestures, and posture. At the margin of our consciousness, although not actually in our awareness, we pick up subtle cues that add information about the relationship and its rules.

I do not like thee, Doctor Fell
The reason why I cannot tell
But this I know and know so well
I do not like thee, Doctor Fell.

Whether it is disliking, as in this nursery rhyme, or liking, we may not know just what it is about a person that touches us. (This is an example of the whole being more than the sum of the parts.)

Through frequent momentary shuttling, we shift our awareness back and forth from others to ourselves and back to others. In this way we can better stay in touch with what is actual, here and now, and respond directly. Thus, we avoid being carried away in fantasy and consumed by emotions evoked by our fantasies. Awareness of others can be partially or totally blocked in many ways: when, instead of using our senses to get information, we substitute our imagination for what is actual, not realizing we are doing so; when we imagine others' intentions, reasons, and feelings and in our mind make them "true"; and when we misinterpret words and metaphors and do not ask the other person, "Is this what you mean?" Also, the present often calls forth the past – our history with the person, or with people similar in some way to this person – and takes us out of the reality of the now. We may hear one word and think we know what the other is going to say, or what she or he really means. Sometimes we hear promises where there are none, and worry unnecessarily about suspected threats that were never spoken. Fears rooted in the past propel us into a future teeming with catastrophic expectations.

The awareness of self and of others are two of the five gateways to meaningful interaction and relationship. The awareness–communication model in Figure 1 indicates the linkages between awareness of self, awareness of others, appreciation of differences,[2] contact, and communication.

This conceptual device, conceived by George Brown, is an attempt to convey the interconnected nature of these five components. Each affects the rest: authentic communication heightens awareness of oneself and others just as heightened awareness

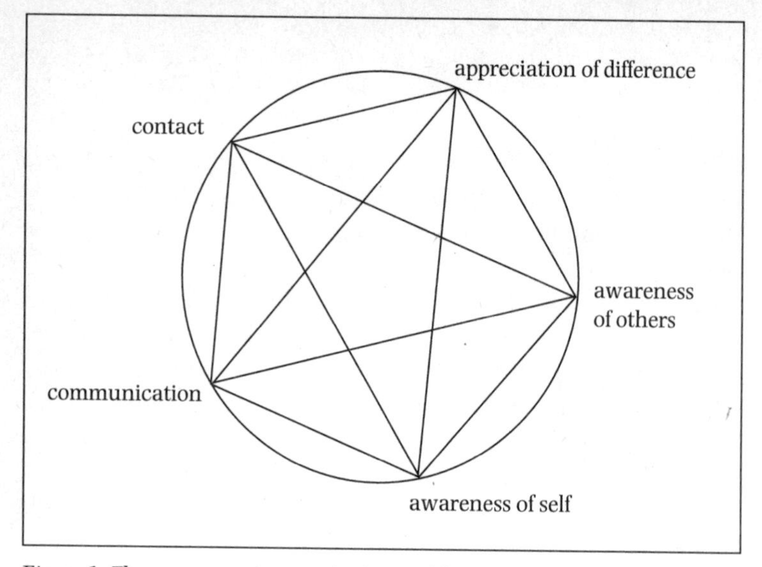

Figure 1. The awareness–communication model

makes authentic communication more possible. Close contact with others brings more contact with ourselves. Engagement and participation in one area spreads to other areas; more awareness of others brings about more contact; more contact opens the door to more self-awareness.

When people are sensitive, their emotions seem to be infectious and intensified through contact.[3] In airports observers can be seen wiping their eyes when long-parted family members greet one another with emotion. While reading historical records a researcher might learn of a tragic incident that befell a person in the study. Later, in relating the information to a colleague, the researcher is surprised by an unexpected sadness, a catch in the voice, and tears. Emotion that was barely noticed in solitude became magnified in company. Something similar may occur after losing someone close to you through death. Eventually you can speak of your loss without tears under most circumstances. Yet you find that telling someone who has not heard of your loss precipitates crying again. We often laugh more, as well, when we share an amusing incident with someone. It may be that in sharing sad news or a joke, we instantaneously project ourselves into the other person's position, identify with him or her in that moment, and react to the sad news or the joke as if we were hearing it for the first time, thus intensifying our reaction. At such moments of shared sensibility, communicated with or without words, we participate in self–other unity, described by Heshusius as "participatory consciousness" (1994:17).

Process focus

Presence, as discussed earlier, rests on seeing and hearing rather than imagining and projecting, being in the moment rather than slipping into the past or jumping ahead to the future. Awareness of and utilization of the present is central to *process focus.* Process focus means attending to what is happening to oneself through direct experience of one's senses and thoughts as one moves through existence moment by moment: "My eyes are almost closing. What's happening with me?" In situations involving more than one person, process focus refers to attending to oneself as well as to the flux of interaction as people move through experiences moment by moment: "I'm getting excited by this discussion. Everyone seems so involved." Process focus assumes, along with attention on the "now," a high degree of self-awareness, the ability "to attend and put words to one's experience, to voice it, preferably in a way that one will be heard" (Papernow 1993:26). In addition, interaction involves "soliciting with curiosity and interest what other people are thinking and feeling about the subject at hand, and staying interested long enough to fully and richly understand the other person's experience" (ibid.:26). Process focus is widely used in counseling, psychotherapy, teaching, and conducting groups. The information gathered forms the basis upon which responses and interventions are then based. "Process focus ... is indispensable and a common denominator to all effective interactional groups" (Irving Yalom, in Lichtenberg 1990:117).

Because it is not judgmental, interpretive, nor evaluative, process focus brings to all interaction the potential for more complete understanding, more significant and powerful contact, and increased equality among the participants. Here is an example: "Several of you haven't participated in this discussion. I'm wondering what is going on with you. Your views are important to me." Process focus involves paying attention to the metalevel and makes explicit what is frequently ignored: sequences of interaction, people's experiences during interaction, relationships, and rules – who makes them, who follows, and who breaks them. Process focus is the maintenance function that accompanies a task. Researchers who have an eye and an ear on the metalevel will appreciate the interconnectedness between individual behaviors. What is more, they can deal with a level of complexity that makes the data richer and gives more meaning and significance to a study.

Metalevel communication

The metalevel concerns a level beyond that of the usual content or subject matter. It refers to the relationship aspect of communication; it is the level where the participants' experiences, their relationship, and the rules and patterns of the relationship are the matters of interest. Paradoxically, to do this one must – momentarily – step out of the relationship to bring about a fresh perspective on the interaction or relationship. Learning how to "step out" is a major component of the course (Chapter 5). *Metacommunication* refers to verbal and non-verbal communication that defines or clarifies the relationship between people who are in communication. It is the relational message that simultaneously comments on communicative processes including the interaction itself. Through metacommunication a researcher can know better what is going on:

> "I'm not sure I understand. Could you say that another way?"
> "I am assuming you mean ... Am I correct?"
> "Your goals here seem to differ from mine. Can we clarify this?"

"From what you said, I infer that you ... Is this what you mean to say?"

"When I ask you about it, you make a face. Do you want to avoid that subject?"

The prefix *meta* makes clear that the communication does not refer to the content being discussed but rather makes explicit what is going on intra- and interpersonally. A group leader makes process comments: "I interrupted you; I don't feel good about that. Please continue what you were saying." The leader's comments communicate metalevel relating, respect for others, and preferred behavior (not interrupting), and they exemplify self-awareness and awareness of others. Messages on the metalevel are self-disclosing; they consist of the interpersonal text, or *metatext*, and may be verbal, as in examples above, or non-verbal. Non-verbal metacommunication could be an intentional look of displeasure or nodding one's head when speaking to someone. It may define context and relationship, as in the seating pattern at a family dining room table or a state dinner. The placing of someone close to the president at a White House dinner makes explicit her or his relationship to the head of state.[4] The degree of formality of dress is another kind of metalevel statement that defines an occasion.

We metacommunicate when we verbally or non-verbally comment on interaction, relationship, and the implicit rules of behavior that define relationship. For example, if a respondent perceives that an interviewer has high status, the former may then make certain gestures that reflect the rules of their relationship – opening doors, offering the most comfortable chair. A person speaking in a commanding tone of voice to a second person implicitly communicates, I am in charge here and you must defer to me. When the second person, through posture and tone of voice, shows deference, she or he metacommunicates, I do not challenge your position. I make myself subordinate to you. Non-verbal metacommunication, ongoing during any interaction between two or more people, can be physical: facial expressions, touching, shaking a scolding finger, or winking. Eye movements, such as eye contact, repeatedly looking away, and slowly lowering and raising one's gaze to take in the other's whole body, further exemplify non-verbal metacommunication. Verbal metacommunication may include

- asking questions ("Why are you telling us what to do?" which contains an implicit challenge);
- answering questions ("I'll let you know the answer to that sometime later," which puts the speaker one up on the person who asked the question);
- interrupting others when they are speaking, which implies a dominant position; and
- using an emotional tone, which may convey dominance, submission, flirtation, or numerous other possibilities.

All behaviors that have the potential to create, define, or clarify a relationship are classified as metacommunication.

Metaprocess[5]

A metaprocess is a deliberate kind of communication with specific characteristics. While more than metacommunication, feedback, and self-disclosure, it includes all of these. An example: "I'm beginning to feel somewhat impatient. It seems every time we start to agree on a plan of action, new information is brought up and we get sidetracked." The speaker begins with self-disclosure: I feel impatient. He then communicates about the current process: we repeatedly get distracted by new information. A request for a change in how the meeting is conducted is implicit in his remarks. The speaker can expect some response, from which he will get more information about others' experiences. Metaprocessing involves processing the process, so to metaprocess one must be or have been part of the interchange or situation. It provides a means to learn about oneself and others in social interactions, and its potential uses are many:

- to unravel and puzzle out the events that have been taking place;
- to clarify the motivations and social dynamics behind the event;
- to reflect on and reveal values, experiences, and observations;
- to allow conflicting views to surface and to highlight relationships, informal rules, and covert alliances;
- to share what has been learned and explicate the steps and operations that facilitated or interfered with the learning or task;
- to show when there has been no closure to the interaction; and
- to create the possibility to come to resolution and closure and to proceed with new pursuits.

While people are expressing different points of view, as is likely during metaprocessing, they may experience new reactions, allow new figure/ground relationships to emerge, and embrace new meanings at a higher level of complexity.

When people involved in a task or study reflect on and process their experience, they process the process. A participant observer or other researcher hears and sees how they talk together, make sense of what transpired, interpret their own and others' behavior, and make connections between events. The researcher may act as facilitator and assist in a metaprocess by contributing the insights and understandings available to one afforded an outside perspective, or metaview, of the system (see below). In addition to shining the spotlight on the group being observed, metaprocessing exemplifies "heuristic inquiry," which "legitimizes and places at the fore [the] personal experiences, reflections, and insights of the researcher" (Patton 1990:72).

Jostein Kleiveland offers this description of a metaprocess: "At the end of the day, the client asked the consultant to chair the meta processing of the day. The consultant asked the participants to divide a piece of paper in half, using the left side to describe what happened and the right side to describe their engagement and feelings. After a few minutes of individual writing, they were asked to share what they had written" (1992:116). Kleiveland then sums up the metaprocessing: "The information obtained from the meta process was that, generally, the group members experienced boredom, anger and frustration, and lack of understanding" (p. 117). Later in the day the client and consultant processed the meeting, including the metaprocessing session, and new realizations emerged from the meta-metaprocess (the process of the metaprocess): "It became clear neither the consultant nor the client had heard what the [group members] had asked for all day.... The consultant and client were so caught up in their own intended programs and hopes for the results, they were not aware of the [group members'] needs" (p. 119).

Merry and Brown emphasize the importance of periodic, planned metaprocessing: "Building in metaprocesses ... allows and encourages continual changes in organizational activities and structures" (1987:282). Kleiveland, referring to the case above, illustrates how such a change is possible: "As a result of the client's and the consultant's perception of their behavior, [through metacommunication during the meta-metaprocessing] and acceptance that their way of manipulating the environment was to satisfy their needs, the client and the consultant had an opportunity to now change the system of how the meeting was chaired and possible response of the participants, which occurrence is defined by Watzlawick et al. (1974) as 'second order change' " (Kleiveland 1992:122).

Metaview

The viewpoint from which we observe determines what we see. A metaview is a view from a distance, a vantage point that allows a broader scope of a situation. It is similar to the perspective of a person joining a meeting that has been in progress for some time. The newcomer remarks on entering the room, "We need some fresh air in here." Those who have been in the room have not noticed. Or, the newcomer might immediately alert the group to a fundamental problem: "Everyone looks exhausted.

Maybe that's why you're not getting anywhere." When we want to know more about our behavior or a situation in which we are involved, we may ask an outsider, knowing that he or she has a metaview: "Tell me what you saw me doing." When one is part of a system, such as a family, a research group, or an organization, proximity narrows one's view. "Knowing" the people and the situation generally confines one to a single frame through which to view the interaction. Familiarity with a system may be helpful in dealing with certain problems and maintaining the status quo, but it interferes with mentally stepping outside the system and seeing a larger view, including oneself in the situation. A metaview provides potential for a shift in framing, increased options of behavior, and changes in relationships.

Metaposition

A metaposition provides a metaview. It is an orientation, a state or stance, that is intentionally initiated or may come unbidden, bringing a "eureka" experience; it may be momentary or of longer duration. It involves

1. *a break from the ongoing process.* This implies stepping outside the system of which we are part; it could be simply a two-person system, or a small or large group. Systems have rules and understandings; we tend to limit our freedom of thought and imagination under their influence. When we figuratively leave the system, we can free ourselves of its limitations. We let go of the content of the interaction, the relationships, and any particular desired outcome or other emotional involvement. Then we are able to take a metaposition. The break may be a split-second shift of attention while the interaction is continuing, or a lengthy period after the meeting or interview has ended when one is reflecting, walking back to one's office, driving home, or perhaps writing up notes.
2. *relinquishing the mental set of the interaction.* Our concerns during interaction vary from situation to situation. We may be tense, moved, inspired, provoked, bored, or eager to push

through our own agenda. As we detach from the content, we step outside all intentions, problem solving, figuring out, and efforts to control or take charge.

3. *taking a metaview of ourselves.* Without trying, evaluating, or judging, we simply observe ourselves: accompanying *physical* sensations, such as rate and depth of breathing, muscle tension, and level of comfort or discomfort; our *mental* state, such as clarity, confusion, or active or quiet mind; and our *emotional* state, such as excitement, anger, anxiety, or boredom. We do this by abandoning our habitual critical judgments, beliefs regarding what kinds of people we are, and convictions of how we should be, simply becoming our own observers. By alleviating our judgments we drop the inevitable conflict of self-accusation and defense. The period of withdrawal, even when momentary, allows a chance to reflect on our intentions and interactions in the time just past. We withdraw our focus to an inner place *without expectation of what should happen.*

What *might* happen when we extricate ourselves from a system in which we are caught is that we find the mental "space" to go beyond our usual constructions. People often cite this likelihood as a reason for taking holidays, for getting away from the workplace and home: "I want a chance to get away and take another view, to think, to get some new ideas." When we figuratively wipe the slate clean, we realize it becomes possible for things to fall into place and for clarity to emerge. A different figure/ground relationship becomes accessible. With this shift come novel perspectives and added knowledge and complexity.

Metaprocess goes hand in hand with and strengthens relationship and communication. Self-awareness and awareness of others are obviously inextricably linked to all interaction. The same psychological mechanisms that impede self-awareness by causing boundary confusion must also hinder our contact with and awareness of others. We further block out others through our interpretations of them and with our judgments and evaluations of their behavior. Here is an example of projection and judgment: "He came late for the interview. He doesn't want to be involved in this. I can see he doesn't respect what I am doing." Listening to our thoughts provides an excellent opportunity for taking a metaposition. For descriptions of ways to facilitate this process, see Chapter 5.

On those occasions when we are (as much as possible) free of all impediments to contact – aware of ourselves and others – we may come to appreciate the differences between ourselves and others, to be capable of acknowledging, respecting, and maybe even valuing them, for we need not be threatened by the fact that the other is truly *other*. We stand face to face as two individuals – separate, different, and safe. Contact and communication are possible. Perhaps only then can we listen with interest, with a mental and emotional willingness to understand the *meanings* of the other's verbal and non-verbal communication.

Essentially, communication conveys meanings from one person to another. We communicate more easily with those who share our language, culture, outlook, and life experiences. Still, difficulties arise when communicating even with them, for we personally construct knowledge on the basis of experience and internalize the meaning of the experience. Words are abstract, loaded with symbolism, and we use them to create metaphors. Useful and poetic as metaphors are, they have particular meanings for the creator and perhaps different meanings for others. Personal meanings are often removed from experience; they depend on background and context. Therefore, they may be difficult for the speaker to express and for the listener to receive. Since contexts and relationships change continually, systematic processing of experience provides a framework for reflection on that experience, which leads to internalization of its meaning (Cook and Caston 1991). If we are to understand one another's meanings, regular communication must be interspersed with metacommunication. Then we need not guess the significance of another's verbal and non-verbal communication. We can verify the substance of messages. We can consider the *mode* of a message when we are not certain – whether the other speaks in earnest or jokes, is serious or sarcastic, straight or ironic.

Summary

The topic of communication claims a prominent place in this work. When one considers the central position of the interview as a source of data in social science inquiry and the interviewer as, one would hope, a reliable instrument for gathering the data, the importance of communication becomes obvious. Communication has two main aspects: 1) the content of verbal and nonverbal interchanges, and 2) the communicators themselves and their relationship. I have earlier established the importance of self-awareness of the researcher. In no context is this more crucial than in interviews, where the interviewer and interviewee simultaneously influence one another. Whether there exists good contact and trust or distance and mistrust, the interviewer must be in touch with what is. The Gestalt principles of awareness, here-and-now focus, presence, and responsiveness are particularly relevant here. Of importance, also, is the ability to achieve a broad view of interactional situations by taking a metaposition.

Metacommunication and metaprocessing can enhance and intensify interaction between individuals and within groups. The ability to leave the level of the subject matter to take a metaview of the interaction, and of oneself as part of the interaction, can lead to a fresh burst of energy in the participants and new strides in completing tasks. Periodic, planned metaprocessing helps to alleviate impasse and encourages change in relationships and structures. This ultimately provides quality data that might otherwise be unavailable.

The learning of communication and metacommunication skills figures prominently in the next two chapters.

Chapter 5

Course Description

We can choose to remain in relative equilibrium by rejecting all input which does not match our cognitive map, and by doing this we choose to become a relatively closed system. Or, as in the case with creative scientists and artists, we can be puzzled, intrigued, or simply frustrated, by complex events or unusual experiences and seek to integrate them.

A. Montuori (1992:198)

It may be conventional to relegate a body of material as extensive and detailed as a course description to an appendix. However, in this book, the course is central to the book's existence. It responds directly to several crucial questions: Is it possible to make subjectivity the subject of a course? If so, can this be accomplished while accounting for individual differences in a personal way? Can the experiences provided as parts of this course be consistently applied in the context of the role of researcher? A related question: Can awareness training for the role of researcher be distinguished from what is commonly thought of as psychotherapy? Can a strong affective component be combined with theory and other cognitive material to make the conducting of research significantly more effective? And can the learning opportunities in such a course be demonstrated to make a worthwhile contribution to the behaviors of its participants in subsequent research roles?

These and many other questions may be pondered in the light of the course content and teaching methodology described in this chapter and Chapter 6. In Chapter 7 some student comments regarding the effectiveness and practicality of their learning are presented to respond, in part, to the last question above.

The foundations of the course at the Institute for Educational Research were developed over many years as we (George Brown and I) taught various groups, including health professionals, educators, and members of military and industrial organizations. We merged basic theory and skills of Gestalt therapy with the experiential approach to the learning of confluent education, both of which are described in Chapter 3. Our broad aims, in addition to presenting the theoretical basis for the course, have been to provide experiences in self-awareness, awareness of others, contact with self and others, and communication skills, as well as to acquaint the students with process orientation. Because of time limitations at the institute, our course was designed to be taught over two intensive weeks. It could also be arranged for a university quarter or semester system or as part of a sequence of methods courses.

I embark on this detailed description of the course with the hope that the reader will take it as an example of our approach and not as the only way to conceptualize training for researchers that includes affective dimensions. This model of teaching suits us and our long experience in process awareness training; more important, it appears to accomplish the broad objectives of the course. We hope, above all else, to stimulate interest, to evoke the reader's creativity, and to inspire experimentation in educating the whole person of the researcher.

Affective learning

A major focus of this course for students who are, or will be, conducting research deals with their subjectivity. Their attitudes, emotions, and values constitute much of the subject matter. Contrary to the conventional university course in which attention is directed to a field of study involving primarily the cognitive domain, the course is designed to engage the students subjectively and thereby make learning personally meaningful, personally revealing, and, at the same time, directly applicable to research. The experiences in the course are planned to further the development of abilities necessary "to meet the demands of self-full ... research.... This requires introspection, reflection, self-awareness and mutuality, empathy, and the capacity to share" (Mirvis and Louis 1988:232). In addition to being valuable for increasing self-knowledge, the methods use and exemplify a holistic view and process orientation fundamental for those employing qualitative research methodology. (Qualitative methodology is specifically mentioned since this course is tailored for those students who will be using qualitative approaches to research.)

Qualities, capacities, and capabilities

There appears to be general understanding about the qualities and personal and interpersonal skills necessary for carrying out sound qualitative studies. In addition to those listed above, Ely, as mentioned in Chapter 2, specified "flexibility, humor, accepting ambiguity, accepting one's emotions" and "empathy" (1991: 113). The list gets longer still with the capacities necessary to carry out Patton's ten themes (see Chapter 3), such as empathic neutrality, flexibility, and the ability to make personal contact. Lowman added "awareness of and control over one's own personal defenses" (1988:183). All lists seem to include openness, and I add yet more: the ability to deal with what is actual, to communicate directly, to be present, to deal with ambiguity, to move into new experience, and to take personal responsibility (that is, to own one's feelings, thoughts, and behavior). Each writer has specific research tasks in mind that call for some or all of these qualities, capacities, and capabilities. By challenging the students with a variety of experiences, we give them an opportunity to explore and discover more of themselves in these areas.

Readiness

Because the learning involved is personally oriented, students may experience a degree of anxiety in anticipation of such a course. Some students will be more prepared than others and more sure of their ability to handle whatever situations might arise. When we consider the ideas expressed in the Montuori quotation above, we recognize how the students are both eager to learn new approaches and at the same time determined to maintain their equilibrium. We pay attention to their degree of readiness, not only during the first hours and days, but during the entire course as each new experience is introduced. We begin gradually to engage the students to the extent of their readiness. The assigned readings repeatedly stress the necessity for self-scrutiny, which supports an acceptance of and eagerness for the work. However, negative attitudes and reluctance toward self-awareness are to be expected, and we find it important to bring such reactions into the open to be dealt with explicitly. Two methods are used to help students monitor their reactions to the activities and to become aware of any difficulties that may arise. The first is keeping a journal; the second is metaprocessing in small groups of three or four students and with the entire class. Both will be described in detail in this chapter.

Classroom environment

The classroom atmosphere and the teacher–student and student–student interactions deserve attention as integral aspects of the course. The climate is accepting and supportive. The values exemplified and explicitly fostered in this course include personal responsibility, awareness of self and others, appreciation of differences, respect for others' experiences, and the capacity to share and to have empathy – all of which are related to interview situations and other research environments.

The phrase *appreciation of differences* (as described in Chapter 4) connotes an awareness that others' experiences may differ from our own, yet their uniqueness is legitimate and valid. An integral part of our teaching is telling stories, usually humorous, to illustrate themes. The following is a story we tell to exemplify the legitimacy of different experiences, different points of view, and the habit of being self-referential.

The story

In the olden days in Persia, the caliph ruled that every village, no matter how small, would have its own judge. In one village an old man, simple but wise, was chosen to be the judge. The village was poor. There was no such place as a courthouse, so the old man used his own house to perform his new duties. The first case that came before him involved two men locked in an argument. He sat them down and said to the first man, "Tell me your story," whereupon the first man told his story. When he was finished the old man said, "You're right. You're absolutely right." Then he asked the second man, "Tell me your story." The second man told his story, and when he was finished, the old and wise man said, "You're right. You're absolutely right."

Well, it seems the old man's wife, very curious to hear what was going on, had been listening outside the door. When she heard what her husband had said to the two men, she was very upset. She rushed into the room where her husband sat with the two men and confronted him. "What kind of a judge are you? This man told his story and you told him he was right, he was absolutely right. Then this other man told his story and you told him he was also right, absolutely right. They can't both be absolutely right." Her husband, the simple but wise man, looked at her and said, "You're right. You're absolutely right."

This story sets the non-judgmental tone of the class and typifies the use of humor.

A participatory, experiential mode predominates throughout the course. The classroom atmosphere is informal, relaxed, and non-critical. As instructors we are perceived as friendly and welcoming toward the participants, responding to their questions concerning both content of discussions and process, accepting diverse views and emotional expression. During all activities the students are reminded to become aware when they are being self-critical, making judgments about how they should act and how they should feel. We suggest they drop self-judgments and be open to whatever emerges for them, to be as a researcher – open and interested without trying to manipulate or change what is. "Right" and "wrong," "good" and "bad," and similar evaluations tend to take students out of their experience and encourage a defensive posture. Rationalizations, justifications, and explanations quickly follow negative judgments, decreasing rather than increasing awareness. Thus they are counterproductive.

It is logical, and psychological, that the methods used in training should match the approach being taught. In this course we make the acts of teaching and learning inseparable from the material being taught. Pieter Lems, in the role of research observer, described our approach in similar training we conducted for chief executive officers: "Principles and concepts introduced as relevant to the training [are] applied to all activities by both trainers and participants" (1989:251). Lems added, "The ... interactions between the [Browns] and the participants also appear to show an absence of any separation between the material [being] taught and learned and the actual process of the teaching and learning" (p. 251).

During class sessions our exchanges with the students resemble those in which researchers use qualitative methodology to gather information. In both approaches, Gestalt and qualitative, a primary goal is to discover what is going on. Lems wrote, "Through the use of the dynamic of *processing,* with the added aspect of using 'I statements,' as well as the emphasis on the use of the meta-position as a tool in framing and reframing individually held positions, the importance of the individual experience and the overarching significance of the continuous flow of awareness as a wellspring for information, throughout the accompanying gestalt formation, [is] created and reinforced" (p. 213). Although there are differences in contexts between the course for business executives that Lems observed and the present course, remarks from students, both in class and in written evaluations, are consistent with the above quotations.

Students' comments and questions are used as springboards for the instructors to facilitate their learning, heighten the experience of their process, elicit more information, and evoke further thought. Expressions of emotions, for example feelings of anxiety in anticipating what we might do or what they might be asked to do, are accepted without judgment.

The students' values and behavior in their current and subsequent research relationships are, apparently, strongly influenced by the quality of relationships experienced in the course. In post-course evaluations, several students have reported positive changes in the way they work and in the manner in which they relate to their colleagues, students, and research participators. These changes suggest the effect of isomorphism; that is, the form and structure of students' work outside the classroom come to resemble the form and structure encountered in the class. Themes dealt with in the classroom setting – accountability, personal awareness, contact with others, trust, power and authority – are relevant to conducting research; further, issues that arise in class are likely to be identical to those encountered in groups and organizations being studied.

The focus during class meetings shifts in a non-sequential way among 1) intellectual work, that is, ideas and theories from the readings that form the basis of the classroom activities; 2) personal experience of the participants; and 3) the relationships between the students and between students and instructors. Sometimes a student directs the attention of the class to a question or problem in any one of the three categories; at other times, we do so. The structure of the class provides time for emergence and flow of focus. We allow, indeed encourage, the students to bring forth their questions and concerns. Because they become aware of themselves and one another in a non-judgmental way, an atmosphere of trust and self-disclosure is created rather quickly. This was exemplified by the parting of a student who had announced at the beginning of the first meeting of the first course that he could attend for only three days. Three days later, when it came time for him to say good-bye, he was visibly moved and saddened to leave; he had become an integral part of the group.[1]

The rapid formation of the close-knit group may be attributed in part to our helping create an open, accepting atmosphere in the class. It is clear that differences are accepted, emotions acknowledged, and contact encouraged. Students feel safe to disagree with one another and to challenge us. The following quotation from a student's post-course evaluation is revealing:

> What really stands out for me when looking back to this course is my experience of being a person with some influence in the group. I never had that feeling in a group before, only for moments, but not in such a way that I experienced some sort of a more accepted position in a group.... I have reflected how this came about. In groups like this I always start with a feeling that I will be happy if I survive at all. So this time too. But gradually it was safe to say things. Even saying controversial things. This especially of George when I said I felt he betrayed Judith in one of [their] dialogues. I met his eyes afterwards and saw he really didn't have any angry feelings towards me.

The frequent formation of small groups for training, metaprocessing, and journal reading undoubtedly increases the understanding, appreciation, and empathy among the students. The high degree of self-disclosure when students engage in Gestalt interaction during chair work (see "The content," below) above all serves to create closeness. In turn, the feelings of care and connection make the sharing and intense work possible.

Since metaprocessing plays a major role in this course, it deserves further elaboration at this point. It consists of communication about people's experiences, their relationships, and the rules and patterns of those relationships. In the present context it refers to the extra time taken after each experience and exercise to discuss what was happening to the students as they engaged in the process just concluded. This may require some moments of reflection. The focus is not on the content of what has just transpired, but rather on the reactions, the emotions that emerged, the feelings that remain. It is a time to check out what was happening for the others, listen as others self-disclose, and become more aware of empathy or lack of empathy – always without judgment – simply to discover more about oneself and others in relationship. For most students this is new and somewhat difficult. Therefore, we provide some supervision to point out instances when they slip out of the metalevel to the content level, or into a self-referential stance, or into a blaming and defending mode. In this way they soon learn to discriminate between the two levels of communication and different modes of interaction. The continual practice of metaprocessing becomes habitual; if we disregard it, either inadvertently or because of shortage of time, we have found the students unwilling to budge from their groups because they themselves need to metaprocess for successful closure of an experience.

As we use the principles of Gestalt pedagogy in our teaching, we aim to make them, as well as the philosophical assumptions, explicit. In this way they are learned, not as an intellectual exercise, but as an integral part of the students' experience. We believe this increases the likelihood that they will apply this learning to their research, teaching, or other endeavors. For example, when we suggest during the latter part of a class day that the students reflect on the day and include in their free-writing any topic or experience that feels unfinished for them, we might ask further questions if time allows: "What emerges into figure from the ground of all of the day's activities? Had you been aware of feeling unfinished in any respect before we asked the question? What do you need to feel complete?" Using such simple questions, we illustrate the concepts of unfinished business, figure/ground, Gestalt formation, awareness of emerging needs, and lack of closure, as well as the experience of coming to closure on a particular issue. We bring depth to experiences by encouraging reflection. For example, after a morning journal-reading session, we might suggest the students think about what they and their classmates read: "What was going on? What were the premises you made? What were your emotional responses to the others as they read from their journals? Does this suggest anything about you in the research situation?" And we connect the activities to research. In the case of the journal reading, we might tell them, "This is the kind of thing you would do when you do research. In your journal you note what your reactions are when you are collecting data. Also, as you reread your journal, you take a metaposition and note your reactions."

The content

The course has two important emphases: process awareness work and process awareness training. The former consists of 1) optional individual sessions in the class with the instructors, the focus being on awareness of moment-by-moment experience, referred to as "chair work"; 2) interaction among group members highlighted by feedback from other students and the instructors to increase the clarity of *intrapersonal* process during the *interpersonal* exchanges; and 3) specific exercises (planned experiences) designed to make more explicit certain feelings and behaviors that may have previously been implicit. During process awareness training, the students learn basic tools of the Gestalt approach. They participate in laboratory experience to hone the skills of seeing, hearing, personal engagement, eliciting meaningful information, and giving descriptive feedback. This is described in detail in Chapter 6.

First day

The descriptions given here of the course and students' reactions are from courses held in September 1992, May and September 1993, and May 1994.

Before the students arrive the tables are removed from the room, except for two or three that are positioned along the walls for students to place their belongings, coffee, and so on. We arrange the chairs in a circle and write on the blackboard an excerpt from Berg and Smith: "The self-scrutiny process is difficult and complex precisely because *both* the researcher and the 'researched' are simultaneously influencing each other. Since this is occurring in ways that initially are out of the awareness of the parties involved, scrutiny is an absolutely necessary part of social science research" (1988:31). Each day's reading assignment is written on the board, as is the awareness–communication model showing the interconnection of self-awareness, awareness of others, appreciation of differences, contact, and communication (see Chapter 4).

As students enter the classroom they greet us and one another. When all have arrived we announce that Anton Hoëm has requested that English be used exclusively in class and for all writing, such as journals. George and I introduce ourselves. We are relaxed and informal. We tell a little about ourselves, mentioning how we got into this work, our more than twenty years of conducting courses in Norway, the Confluent Education program, and my private practice. In each course there have been one or two students whom we had met before.

Free-writing

From the first minutes of the first class, the students are encouraged to become more self-aware. They themselves are a major part of the subject matter of the class, and we make this explicit when we ask them, upon concluding our opening remarks, to free-write for three to five minutes about their expectations for the course, their apprehensions, and anything else that comes to them as they write. We explain what we mean by *free-writing*, stating that they will not be asked to share what they have written if they prefer not to.

Rationale

From time to time during class, we ask the students to free-write for three to five minutes. Free-writing means spontaneously recording notes of whatever comes to mind without censoring or worrying about proper sentence structure, grammar, or a potential reader. This activity is used to highlight an experience while it is still vivid, before time has elapsed and subsequent events cloud impressions. Taking a few minutes between activities or discussions to jot down ideas and reactions helps one remember and integrate the material; it contributes to making connections between what has just transpired and similar or different experiences from the past. The entries from free-writing are incorporated into the students' journals.

In asking them to free-write we put into operation some of the practices we will be stressing throughout the course: self-awareness, reflection on what has just occurred, taking a metaperspective – all aspects of paying attention. During our two weeks together we hope to rectify habits of inattention, unawareness of one's reactions to experiences, and deficiencies in seeing and hearing.

Assigning free-writing at this time introduces

- connecting with the experience of the present moment;
- putting into words and writing down their thoughts and experiences regarding, for example, what they may have previously heard about us and the class, or what has taken place for them so far in the first session; and
- taking a metaperspective.

Having asked the students to free-write, we could follow up by suggesting they read over what they have just written; also, we could pose questions to help them focus on their experiences of writing, their reactions to what they included or excluded, and their attitudes as they read what they wrote. Do they find they were judgmental of themselves in what they wrote? What did they neglect to write? Introducing such ideas would acquaint them with the practice of taking a metaview (actually a meta-metaview) of the experience of free-writing, which is itself a meta-experience.

We could also build on the experience by having the students get together in small groups to discuss what they have discovered, to metacommunicate about what has happened for them since they arrived in the classroom. This would be aimed directly at developing "mutuality, empathy, the capacity to share," further requirements of qualitative research mentioned by Mirvis and Louis (Berg and Smith 1988:232). However, we choose to forgo such reflective experiences at this point in order to get on with the content level of the class. The question of prudent time allocation is always a consideration for us. At those times when we are dealing primarily with content, we attempt to balance the instruction with process comments and questions; when we focus on process, we include some theory.

Introductions

George explains to the students how we would like them to introduce themselves. From their places in the circle, they will come and sit in a chair that is placed between us and, along with their names, tell anything else they would like us and the other students to know about them. George motions to the person next to him and asks him or her to begin. As the first person takes the seat, I address the group: "I want to ask you to become aware now of what happens when George says we want each of you to come up to the chair here and introduce yourself. Do you have a different experience in this moment?" I notice some smiles and nodding. "Do you begin to rehearse," I continue, "or plan what you will say when it is your turn? Are you aware of any physical sensations? Please simply become aware of what you are doing and what you experience. If you are judging yourself or criticizing, become aware of that. Perhaps you can drop your judgments and simply notice how you do that. If you are rehearsing now, or begin to as your turn comes closer, recognize that is what you are doing. At the moment you become aware, you have choices: to continue to rehearse, or to stop rehearsing and listen to what others are saying when they are in the chair. You may find you do some of both. Perhaps you can trust that you will say whatever you have to say when your time comes, in which case there is no need to rehearse."

We invite each person in turn to come to the chair. They tell about their work, their research, their families, experiences they may have had with the Gestalt approach on other occasions, or perhaps how they heard of us. Some tell how they anticipate the coming two weeks we will be together.

Rationale

Asking each student to come out of the circle and take the chair to introduce herself or himself represents a significant first step in making explicit our expectation for active participation in the class. Our questions about their experiences in the *now* (e.g. "Are you rehearsing?") introduce a major focus that we will return to again and again – their subjective experience in the moment. We present the idea that with awareness comes choice. This form of introduction offers an opportunity for self-awareness, awareness of the others as they introduce themselves, a recognition of their individual differences in manner and reactions, and feelings of connection with others. In inviting the students to come to the chair at the front of the class, we ask for a commitment of involvement. In coming forward, they make that act of commitment.

After the introductions we ask the students to free-write about their experiences when they were in front of the others to introduce themselves and when they observed the others talking about themselves. Then we take a short time for metaprocessing in the large group.

Metaprocessing

Sometimes called simply "processing," metaprocessing is the process of discussing one's reactions to what is, and/or has been, going on (see Chapter 4). Metaprocessing combines feelings and thoughts: affective components along with the cognitive. Reflection may be spontaneous or deliberate. It requires a break in contact with the situation to separate emotionally from ongoing activity or conversation to get a metaview. The time taken for withdrawal may be a fraction of a second or some minutes' disengagement from the content of an interaction and the emotional involvement that entails. Taking a metaview serves many functions: it provides a broader perspective of what is happening; enables people to view their own participation in creating what is, or has been, going on; and allows appreciation and understanding of the experiences of others. Metaprocessing is built into every activity during the course, both to sharpen the skills of the students in this difficult practice and to further awareness and engagement in the class.

The relevance of metaprocessing for working with individuals or groups cannot be overemphasized. Conscious or unconscious withdrawal from any focus of attention, even for a moment, permits the surfacing of impressions, thoughts, and feelings that had been out of one's awareness or quickly pushed aside during the concentration on an activity. In group process, when many events happen simultaneously, a chance to reflect on what has just transpired can bring a metaperspective, sudden clarity, and reframing.

After metaprocessing we talk about the course in general. George explains the awareness–communication conceptual model (awareness of self, awareness of others, appreciation of differences, contact, and communication; see Chapter 4), which is written on the board. I present our general plan for the succeeding days (see p. 68).

After a short break we do the first of the group Gestalt exercises: Suffer/Terrific.

Suffer/Terrific

George introduces the Suffer/Terrific exercise. He explains that people coming into a new group often have a desire to be noticed, to stand out from the others in some way. At the same time a new participant may want to be anonymous, just one of the many, no different from everyone else. Two ways to be noticed in a group are commonly used. One is to complain about something; the other is to boast, to let people know you are someone special. Now, George tells the students, they will have a chance to do both. He asks them to form pairs. One person in each pair tells how she or he suffers while the partner listens and encourages the speaker to say more by repeating, "Tell me how you suffer." The asker is confined to this one sentence. When each has had a chance to play both roles, they metaprocess this first part, then proceed to the second part of the exercise: "Tell me how terrific you are." One partner tells how terrific she or he is while the partner repeatedly asks, "Tell me how terrific you are," to spur on the one who is boasting. Again, they switch roles. When they have each had a chance to play both, George asks them to process the exercise by talking together about what they experienced during this activity, both when they were speaking and when their partners were speaking. Did they find it easier to speak about suffering or about being terrific? During the processing George asks, "How could the attitude you had when you told how terrific you are affect you in your work?" With that question they seem to change the level of communication, to personalize it and thereby deepen it.

We return to the large circle and ask whether there are comments or questions. A student comments, "I felt empathy from him, so it was easier to boast." I ask, "When interviewing, when might you have empathy? When might you lose it?" A discussion follows: There can be danger in too much empathy; one then risks becoming part of the other's system, losing the ability to maintain a view of non-attachment, a metaview. Processing continues in the large group. Students add their comments: "It would be useful to apply this experience to my work." "It was demanding to engage in this situation – to speak and to listen." I ask, "How were you when you were listening?" A student replies, "I identified with my partner's boasting."

Rationale

Until the Suffer/Terrific experience, the students have been in the large group. The act of pairing up, choosing a partner or being chosen as a partner, requires a different kind of participation; each student must carry the weight of being half of a couple. In this situation people will say things they might not want to share with the whole group. This exercise is not too risky; therefore, it is a good one to start with. (Although we use the word "suffer" students may complain about anything from "metagrumbles" [Maslow 1971] to more serious life situations.)

There is generally an element of humor, but the exercise is also taken seriously, and students are willing to self-disclose to some extent.

Through this exercise we invite the students to tell something about themselves. Using this particular approach avoids the temptation to take an easy path and engage in small talk. This format also equalizes the relationships by enabling each partner to have equal time.

We take this opportunity to mention paradox for the first time – specifically, the "paradox of belonging" (Smith and Berg 1987). It is usual for people to have contradictory reactions when joining a group. Each wants to be a member, to be accepted as part of the group, to have some influence, yet at the same time each wants to maintain his or her individuality, independence, and privacy. By bringing up this concept we allow students to acknowledge their own and others' cautious attitudes in new experiences, without blame or negative judgment, and simply as "what is" for them at the moment. Paradoxically, change occurs through the acceptance of our experience rather than attempts to change it. At times during the course we again discuss paradox in relation to intra- and interpersonal relations and the process of change.

After a lunch break, in introducing the afternoon's activities, I speak of the idea of increasing commitment by taking small steps. For example, the students' act of coming up to the chair to introduce themselves at the beginning of the class perhaps made it easier for them to make a fuller commitment to participate in the subsequent experiences than if they had stayed in the circle and spoken from there. I explain that the exercises we provide are *purposeful*, although no specific outcomes are expected. We invite them to participate in the activities to the degree they choose; the experiential portions of the class work are opportunities for them to discover more about themselves. We remind them to become aware regarding whether they are making judgments about themselves and how they should act or feel. If they are doing so, we suggest they drop their judgments when they become aware they are being self-critical and simply discover whatever else emerges during the process of the activities.

Demonstration

Each time we introduce a new experience, we demonstrate what we are asking the students to do. For example, when we ask them to become tarot cards (see below), one of us will take a card and demonstrate. We try to be as spontaneous as possible, purposely not saying things we have said during previous demonstrations in other groups. If we ask them to pair up and sit facing one another to practice giving non-judgmental, descriptive feedback (as we do in the process awareness training and other experiences), we first sit opposite one another and demonstrate. After our demonstration we ourselves take time to metaprocess our experience. We have found, and heard from participants, that these short demonstrations are most important. It seems our own spontaneity, authenticity, and willingness to self-disclose in the moment set an example for what is possible. We do not, at any time, suggest they should be like us. Rather, we emphasize that they can be only who they are and involve themselves to the degree they are inclined.[2]

Throughout the course as the students observe our demonstrations, they come to experience us less as professors and more as human beings. They also discover the potential depth and excitement of a simple, formularized exercise if they care to take it seriously as a learning experience. Perls described learning as discovering that something is possible, and we hope they discover what is possible for them in affective learning. Lems observed, "No behaviors appeared to be demonstrated as such, rather the [Browns], by avoiding using their behaviors and actions as a model per se, allowed for the participants' individual experiences to develop and be recognized as legitimate in this environment" (1989:4).

Tarot cards

Tarot cards are the forerunner of present-day playing cards – an ancient tool for reading fortunes. There are various decks. We use the Waite/Rider deck because the pictures seem most useful for working with projections, for taking a metaposition and reframing. All cards in the deck are picture cards. Most show people in nature, individually or in groups, in various activities and situations, such as a knight riding a horse across a landscape, or a couple with some children standing under a rainbow with a castle in the background. The scenes may be interpreted in innumerable ways, depending on the person looking at the card. We instruct the students to choose two cards from the deck, which has been spread out face up on a table. They are to take one that they respond to in a positive way and another that is negative for them. In groups of four the students in turn "become" their cards by talking about them as themselves: "I am a knight. I ride through the world on a beautiful horse. My function is to help people." They elaborate as much or as little as they please, not to create a full story, but to describe more fully their existence, situation, and how they experience it. Some cards have no people in them, in which case the student becomes the objects in the card: "I am this card. When you first look at me you see only a cloud and a hand; but as you stay with me longer I remind you there is more to life than what you can easily explain." The student continues in his or her own way.

There are many possible ways to use the tarot cards. We generally include five stages:

1. Talk about the positive card as you ("I am this card. I have ... I am ...).
2. Do the same with the negative card.
3. Holding one card in each hand, carry on a dialogue. What do they say to each other? (First card: "I'm not sure what to say to you. You are a king and you might not listen to me." Second card: "I don't know what to say to you, either. We

have never had a conversation before." And so the conversation continues.)

4. Reflect back on what qualities and characteristics you ascribed to the cards when you first became them ("As the knight I was young, energetic, arrogant, helpful ..."). Consider which ones you recognize as aspects of yourself. Remember to be specific: When? Under what circumstances? With what people? (We suggest they become aware of whether they judge themselves as they do this and suggest they drop their self-

Figure 2. First published in Puck *in 1915.*

assessments temporarily. We also ask them to consider which qualities and characteristics they do not own, that is, do not recognize as part of themselves.)

5. Take the negative card and redefine it as the positive card. Then take the positive and redefine it as negative. (The inclination is to use humor and become a caricature; we encourage them instead to discover qualities that were not in figure before.)

When they finish we ask them to free-write, urging that they be sure to write something about the experience of reframing: whether they were able to do this, and what was necessary to do it. Then they talk together in their groups of four to metaprocess. When all return to the large group, we ask for comments and questions. If there is time we talk about the process of reframing – the necessity to let go of a fixed gestalt and allow a new figure/ground relationship. I may at this time show the gestalt picture of the young and the old woman to illustrate the process of one formation's destructing before another can emerge. In connection with reframing the negative card, I remind them of Jung's caution: If we do not own our shadow we will meet it outside ourselves. I add, "and marry it." We ask them to consider their attitudes toward the others in their group. Were they judgmental? Could they put aside their judgments? Were they empathic?

Rationale

For the first exercise (Suffer/Terrific) we put the students in pairs. For the tarot card exercise we expand the groups to four so that they gradually become more acquainted with one another. In this context students seem to feel more threatened by those they do not know than those with whom they have shared experiences. We assume they are beginning to feel safe enough with us and their fellow students to move out of the purely cognitive to a more experiential realm, and we invite them to take that leap with the cards. When someone begins by saying, "I am this card," they are no longer simply *talking about* it, which is some steps removed from an experience; rather, they are *moving into* a new experience. To some degree many of them surrender to whatever emerges moment by moment. Others may manipulate the experience to assure their control. Either way they share the experience with others and discover how they are and what they need in this situation.

Using tarot cards is an efficient way to deal with self-awareness and to encounter projection, both in oneself and in others. Some cards have a distinctly negative cast (e.g. a man lying face down with swords in his back) and are generally chosen as negative cards. However, sometimes they are chosen as positive cards, which may surprise others in the group. Many cards are more neutral. The participants observe one another's cards and realize how large a part each person's subjectivity plays in choosing cards and in speaking as a card. "I would never have thought to have said that" is a common reaction upon hearing others in the group. When working with large classes we use two decks; on such occasions it sometimes occurs that two people in the same small group will find themselves with identical cards. The two always have very different things to say about the cards as themselves. It becomes obvious that the greater part of what is said about the cards does not arise from the pictures but is derived from within the person and attributed, or projected onto, the cards.

The cards are used as a form of symbolic representation, a chance for the students to become acquainted with metaphors and analogies in the context of the course. The exercise provides the first of several opportunities to take a metaposition and experience reframing – a shift in figure/ground relationship. Equally important is the encounter with experiential learning and the engagement, along with others, in a process of discovery and discussion: the opportunity to talk together about an event in a safe, non-judgmental environment. In addition to the intrapersonal discoveries, there are interpersonal dynamics. This experience with the tarot cards is a strong one for many of the students and enlarges the foundation for the practice of ongoing self-scrutiny and the expansion of self-knowledge.

We have the opportunity to listen in on the groups and give feedback from time to time. We encourage them by saying, for example, "You are talking *about* your card; can you *become* it instead?" At other times we might say, "Don't judge yourself about this. We don't give report cards. Simply become aware." I once noticed a group consisting of three women and one man; the man sat back in his chair, saying nothing. The women all leaned forward and spoke to him. As I moved closer I heard them explaining to him what he was supposed to do. I described to them what I was seeing and hearing. I asked them what they were doing. One woman laughed a little. "We're explaining what he is supposed to do," said another. "Oh," I said with a smile, "aren't you a lucky man! Three women to enlighten you. How did you manage that?" And we all laughed.

Before ending the class George asks each student to comment about the day. We then detail what we want them to include in the journals they are to keep during the two weeks of the course.

The general plan

Subsequent days commonly follow a set pattern: journal reading in small groups, process awareness work, discussion of the previous day's reading assignment, group Gestalt experience, process awareness training, and finishing up the day. Time is taken for free-writing, and metaprocessing is built into all exercises. It must be emphasized that we are constantly pulled in two directions: we are flexible, taking extra time when needed to deal with what emerges; yet we also have an eye on the clock, wanting to fit in as many of our planned activities as possible. The following is the plan in more detail (times are approximate):

- Journal reading, 9:15–9:40. Mornings begin with the students in groups of three or four reading aloud from their journals. They are requested not to ask questions or give feedback during the reading since this would use more time than we have. If there is extra time after all in a group have read, they may use it for questions and discussion.

- Process awareness work, 9:45–10:15. Whoever would like to comes to the chair that is placed between George and me, and we proceed to focus on awareness and process. The session usually lasts 15–30 minutes.
- Free-writing followed by group metaprocessing of process awareness work, 10:15–10:45.
- Break, 10:45–11:00.
- Discussion of reading assignment from previous day, 11:00–12:00. The students form small discussion groups in which they share ideas, comments, and questions about the readings. Sometimes we listen to the groups and make comments; sometimes the students call one of us over to ask a question. We are careful not to stop the flow of their discussion.
- Break, 12:00–1:30.
- Group Gestalt experience, 1:30–2:30. These are experiences in which all students engage at the same time, sometimes in pairs, sometimes in small groups.
- Process awareness training, 2:30–4:00. In groups of three, students become acquainted with and practice the basic skills of seeing, hearing, putting words to personal response, and using basic questions to heighten awareness (see Chapter 6).
- Free-writing and/or group processing of the day, depending on available time.

The journal

Students are expected to make journal entries after each day's class. They use the journals to keep a record of their day: their thoughts about and reactions to class activities, relationships with other members of the class, responses to each day's assigned readings, and whatever else seems relevant to them. Journal entries are read aloud at the beginning of each morning in small groups. Portions may be kept private, as desired. Journal writing and sharing in small groups aid introspection, reflection, self-awareness, the capacity to share, appreciation of the variety of reactions among the students to the class activities, and development of empathy through listening to others.

Cultivating the habit of keeping a journal is relevant for the students as members of the class and also applicable to the role of researcher. Journal writing while collecting research data provides a regular opportunity to take a metaview of one's attitudes, reactions, and hypotheses. It creates a record that can help clarify negative developments that interfere with optimum functioning, such as confusion between people due to strong emotional reactions, or parallel process, in which one entity – for example, a research team – unconsciously assumes the behavior of another entity – the group they are studying (Berg and Smith 1988:30). (Parallel process has been noted in other situations, including those in which a psychotherapist works with an individual, a couple, or a family.) Strategic decisions need to be recorded at each step of a research project. A journal, or log, should show the emergence of hypotheses as they take shape during a study and how one moves from one step to the next – "Here's where I thought I was going. Now here's where I think I'm going." The log itself becomes a source of important data for the methodology section of a study.

Journal reading

In each course we have asked the students whether they are comfortable having us listen in on the journal reading in the small groups. As the students read we sometimes sit in with a group, with their permission, or go from group to group. Once or twice a student has waved us away. By hearing what they have written we keep in touch with how they interpret and experience the class. We get a general idea of how they are reacting and whether they are having problems with any of the class activities or have misunderstood the reading assignments. In the large group we metaprocess journal reading. The following comments from students reflect the consensus that the time used to keep journals and share the entries is well spent:

- "I like to reflect on myself and to write it down. To read it aloud is good; it gives me a good feeling. Someone is listening."

- "This is very good training. I generally use too much time to write. I think too much. Therefore it is good practice."
- "Both free-writing during class and journal writing in the evening help me to see things I didn't see at the time. A fruitful meta."
- "Making notes during reading requires thought; then it is not just the author's text but almost a dialogue."
- "I learn very much from this awareness work – the metalevel. I get ideas of what is behind what I do. It gives the possibility of hearing *our* rationality for what we do."
- "So many points of view come through when we read the journals."

Further, the group metaprocess can evoke intriguing and stimulating questions:

- "Yes. What happens, happens. We each see it and describe it differently. Is there one full description?"
- "How do you describe a situation if it is emotional? What words can portray that?"
- "Is the situation emotional or is it only your perception that makes it emotional?"

Process awareness work

In all of our interactions with the students, we hope to heighten their awareness of themselves and their environment, to help them learn more about themselves – their primary research instruments – and hone and effectively use these instruments. This objective is fundamental to the course. The process awareness work that we introduce on the second day is an intense interactive process with a moment-by-moment focus between us and a student who volunteers to come to the "working chair." This chair (which we sometimes call the "adventure chair") is placed between George and me; the students' chairs form a horseshoe shape. There are no tables. In front of the working chair there is an extra chair, sometimes called the "projection chair," for making explicit and working with disowned roles (Perls 1973:120).

As a warm-up step we ask them, as they sit in the group, to focus silently on themselves as we direct their awareness. We ask questions such as, "How are you breathing now? Do you change your breathing when you hear this question? Do you experience any tension in your body? Where are you tensing you muscles? What is the expression on your face? Are you talking to yourself?"

We tell them they may come to the chair if and when they elect to do so. They are not required to do anything in this course they do not choose. We elaborate on "personal response-ability" (the ability to respond), a basic principle of the Gestalt approach. At no time do we expect them to surrender their ability to respond to a situation or give up their power to choose what they prefer. They do not need to rehearse what they will do or say. A student may or may not initiate by talking about something that is on his or her mind. What they do in the working chair is neither right nor wrong, but simply what they do, and that is what we focus on. During the Gestalt process we will, to use Perls's phrase, "make what is implicit, explicit and state the obvious"; we give non-judgmental feedback based on what we see and hear and our personal responses. Our task is to focus on the *now,* moment to moment, to help them engage in their process with consciousness. What follows are two brief examples of typical process awareness interactions. In both cases the participating student was one who accepted our invitation to "have an adventure." Our thoughts and hypotheses are in italics.

Student: (looking down) I feel everyone's eyes on me. They make me nervous.

Judith: I'd like to explore with you what is going on. Is that all right for you?

Student: (glancing at Judith) Yes, that's OK. (She glances quickly at the group.)

Judith: I notice you take a quick look at the group.

Student: Yes, but I don't really see them.

[She feels eyes on her but she herself does not see. She projects her eyes out onto the others.]

Judith: Please move to this chair, and now you be the eyes looking at you.

[Is she ready to be the eyes looking, to take back the power she gives the others to judge her? Can she reown her projection?]

Student: (She moves to the extra chair, crosses her arms over her chest, and cocks her head as if to scrutinize herself.) Yes, we are all looking at you. We watch everything you do.

George: (addressing her in the extra chair as "Eyes") Eyes, tell her how you respond as you look at her, as you watch her.

Student: (as "Eyes" looking at her) You seem a little nervous, but we don't mean to make you nervous. You know, we think you are quite brave to be up there. (She looks at Judith, smiles a little.) That's a surprise!

Judith: Come back to this chair now. (She returns to the working chair.)

Student: (She takes a deep breath, then looks up at the group.) They don't seem so scary now.

George: Tell them that directly.

Student: How strange – I see all of you and you've really changed. You don't frighten me now. (The group laughs. She looks at George.) What happened? I don't understand what happened.

During this short demonstration of process work, we highlight how the student avoids using *her* eyes and taking *her* power. She feels victimized by the power she projects onto group. ("They make me feel nervous.") We ask her to go where the power is: to become the eyes of the other students. Her posture in the extra chair at first suggested she would start judging herself harshly, but almost immediately there was a shift of attitude, which seemed to surprise her.

The next example demonstrates a different approach. It begins with the student sitting quietly, shifting her glance from me to George and back again. She says nothing.

Judith: *[My impression is she is waiting to be told what to do. Will she ask us directly for help?]* You look at us and say nothing.

Student: I don't know what to say.

Judith: In this process work we are interested to hear what is happening with you so we might help you get more totally into your experience. When you say you don't know what to say, you are telling us your experience.

Student: (long pause) This is difficult.

Judith: *[She seems to ask for help indirectly. I suspect she does not realize this.]* Yes, I see you are having some difficulty. I suggest you stay with your experience of not knowing what to say, of finding this difficult. (pause) Simply repeat the sentences you just said. If your experience changes, report that. *[I help her and do not help her at the same time. My hypothesis is she is avoiding experiencing not knowing, being unsure; she wants to do the "right thing."]*

Student: You mean just keep saying I don't know what to say?

Judith: You ask me a question. I wonder whether you could answer your question yourself. (Student looks at Judith but says nothing.) *[I imagine she is getting desperate for help.]* Would you move to the other chair, or just stand if you prefer, and see whether you have an answer to your question, "You mean I should just keep saying I don't know what to say?" *[I think she is sinking into total helplessness, with no self-support, yet I am reluctant to tell her what to do; I want to encourage her to do what she can for herself instead of imagining we have to do it for her.]*

Student: (She stands and looks down at the chair where she was sitting. She looks at Judith, who says nothing, but gives a little nod as if to say, It's OK, go ahead. She begins.) I don't know what's wrong with you. You come up here and then do nothing. You have plenty of things to say.

George: Does she have plenty to say? What do you suppose she needs in order to be able to say these things?

Student: Just to open her mouth and say them.

George and Judith: Tell her that.

Student: (in a stronger voice) Just open your mouth and say

what you have to say. (looking toward us) She gets like this and I feel so frustrated.

Judith: You seem to explain to us about her and you. Please come back to this chair and tell the others in the class about this.

Student: (switching to the pronoun "I") I do get frustrated with myself. I know I don't have to be so helpless.

George: Are you good at getting people to help you?

Student: (smiling) Yes.

Judith: Well, today you helped yourself. (Student nods and smiles.)

In this interaction we followed the Gestalt principle of not doing for clients what they can do for themselves. The habit of projecting power out to others in one's environment (or to institutions and organizations) and disenfranchising oneself is what might be called a "neurotic symptom." In both of these sessions the students did just that. This moment-by-moment process work can help both participants and observers to experience what they are doing and how they do it. Then with a little guidance, a little support at the right time, they find what it is they need and in the process find their own voices. They undergo a moment of "taking back," of coming in contact with, some ability they had been projecting. They experience an integration in the moment – a moment of self-actualization. In this way they discover they need not manipulate others to do for them what they can do for themselves. For Perls (1988:28), this is maturity: generating self-support. For researchers we see this process as a way to more finely tune the instruments that are themselves. The more of themselves they have available, the more effective they are. Interactions are most successful when one is clear, direct, and integrated.

As we work with students in the Gestalt interaction, our primary goal is to make the theory of the Gestalt approach into a vital practice of discovery. We focus on the present moment, facilitate their process, and help them become aware of how they engage in and how they interrupt their process of organismic self-regulation and the *process of growth*. Self-regulation refers to Perls's belief that all organisms have needs and naturally behave in ways to meet those needs. By helping the students experience their behavior in simple, non-threatening ways, we find they are often able to make connections with their dominant needs. "Far from being trivial, the uppermost surface of our behavior is, in fact, the expression of our present dominant need and of our unfinished business" (Clarkson and Mackewn 1993:93). For example, if a student falls silent, or seems to go into a partial trance, we might make that explicit: "What is the need that accompanies, or precedes, your behavior?" Such a question does not call for an interpretation, either by us or the student, but rather invites and often evokes a waking-up to what he or she is doing and to what he or she needs at the moment: "I need to withdraw now," or, "I'm thinking about what I should do now." To intensify the experience and make it more complete, we might ask the person to tell us and/or the other students directly, "I need to withdraw now." Because much behavior is automatic, people are unaccustomed to connecting with their needs and discovering whether their behavior brings satisfaction and, if not, whether options are available.

Because process awareness work is based on principles of Gestalt therapy, a natural question would be, "Is this therapy and, if not, how does it differ? Process awareness work is not therapy. We are constantly aware of the differences and take care not to cross the line. How do we draw the line? We keep in mind the *purpose* of our demonstrations: to illustrate examples of process focus. At each intervention we stay within the bounds of "what and how, here and now." We do not initiate topics. We do not work with dreams. We do not follow up obvious leads to deep, unfinished situations. We do not assign homework that would focus on therapeutic issues for certain students. We do not help students identify or solve personal problems. If someone comes to the working chair with a first sentence that sounds like an invitation to therapy, such as, "I realize I have a lot of unfinished business with my ex-wife/father," we do not accept the invitation. Instead, we stay with the actual behavior in the moment to increase awareness of self and/or context. For more on this topic see "Rationale," below.

Students may become aware of painful situations in their personal lives when they observe another student in the working chair or when they themselves are sitting there. In this context it is accepted – and appropriate – to cry when one is sad, as it is to laugh when one is happy. Crying is a natural expression and does not make an experience "therapy." Therapy is generally thought of as a series of sessions in the privacy of a therapist's office or in a group to work on specific problem areas. The usual goals are to alleviate symptomatic behavior and improve one's quality of life. By contrast, our process awareness work usually consists of a daily session, with a different partner each time, lasting 10–30 minutes. Occasionally a student "wakes up," gets more in touch with what is, and takes a step toward maturity. The work may also help students to break out of habitual, repetitive behavior into behavior with awareness and more options. This, however, is not our goal or expectation.

On one occasion, after a student cried in grief over a loss that had occurred nine years before, a student questioned, "How far should people go into their process in the chair?" The student who had cried replied, "I wasn't going into anything that I didn't want to go into." To us she remarked, "Your comments kept me on track. I felt finished when the session ended."

In the spirit of confluent education we provide theory, bringing cognitive dimensions to add meaning to the experience.

Secondary goals of process awareness work include

- discovering the difference between experiencing and figuring out,
- trusting in the process (see Chapter 3),
- expanding the ability to come in contact with others and with oneself,
- experiencing the confusions that can result when we project what is our material onto others,
- focusing on the skills the students will practice in the process awareness training,
- fostering deeper recognition and increased empathy for one another, and

- demonstrating the paradoxical nature of being engaged and able to maintain a metaposition.

After demonstrations we process the work in two stages. First, the students who have been observing tell what they experienced during the work. We always include time for this exchange, because when students engage in process awareness work, they have experiences that touch everyone in the class. It is important that the others have a chance to say something; thus, they achieve some completion and are not left experiencing strong incomplete gestalten (unfinished situations). Second, we attend to any professional questions they may have about the work and its connection with their roles as researchers. Then we take a short time for free-writing.

The comments from the students to the person who had been in the chair typically express *interest* ("I was with you every minute you were in the chair"), *surprise* ("Your work really demonstrated staying with your process and what can happen when you do that"), *involvement* ("I was laughing and crying with you"), and *caring* ("I felt very connected with you while you were up there"). Many identify with the student who comes to the chair in connection with the content he or she brings, for example, fear of facing the class in this way, the burden of carrying unfinished issues from the past that are emotionally laden, and anxiety about the future. They often recognize patterns of self-interruption, which are made more obvious in the work. The following are student comments from various work sessions:

- "It was good for me to see your anger and aggression. Nice to see you come out with your feelings."
- "I very much identified with you during that painful part. And when you seemed resigned, oh, that was terrible for me."
- Student who had been the focus of a session: "I get touched again as you speak." Other student who had observed: "So do I."
- "That was a strong statement you made at the beginning: 'I don't know what to do here.'"

- "When you looked at me and didn't know my reaction, it was right. I was going back and forth, identifying very much at times. The last time, I knew you were irritated with me; I was irritated with myself."
- Student who worked after hearing others respond: "How genuine everyone is in their reactions and comments."

As indicated in the general plan above, we include process awareness work each day so all students who choose to do so have an opportunity to come to the chair.

Rationale

Students gain more awareness of their ongoing process through engaging in and observing the work. They discover how they organize their experience to get what they need in this particular situation, whatever it might be: attention to an unfinished situation, a particular reaction from us or the group, or avoidance of encounter. The power of authentic expression is recognized as natural and not to be feared. Participants and observers alike are reminded of the complex forces that, although generally kept under the surface, exert influence moment by moment. Simple encountering *what is*, making *what is* explicit, intensifying the experience, and moving with one's process combine to constitute a startling new event for most people.

Much of what we bring into focus here has direct relevance for researchers: body posture (both their own and that of those with whom they interact), body movements, facial expressions, and eye movements (i.e. when and if the respondent glances at or away from the interviewer); listening to the voice and manner of speech as well as the content; and differentiating between the experiential mode and thinking about, fantasizing about, figuring out, or trancing out. The students hear how rich with meaning the first sentence spoken can be and how empty the content can be at other times, when the posture and the voice quality may be more meaningful. Contactful interaction observed during chair work, difficult as it may be to describe, is easily contrasted with simply "going through the motions." Further, process awareness work has the potential to demonstrate process focus and trusting the process, both assets in collecting and analyzing data. Not only the student who is working but also the others can discover how the total experience of *what is* allows something new to emerge, and that when one gives up *trying*, clear and meaningful wholes (gestalten) often present themselves.

Group Gestalt experiences

Throughout the history of our work we have used a large assortment of exercises, many of which were created to meet the needs of an individual or group in the moment. Through the use of exaggeration, or by temporarily isolating one aspect of a person's internal process, our objective is to make what is implicit explicit in order to face, acknowledge, and deal with it. In the context of this course the following need to be considered with regard to the choice and timing of exercises: the readiness of the students; the needs of the class (which might dictate, for example, a particular emphasis in the process awareness training); the degree of risk; and the course objectives.

The Suffer/Terrific and tarot card exercises that we included on the first day of class have already been described. Other group exercises (and when we typically include them in the course) are Topdog/Underdog (usually scheduled early in the course); See, Imagine, Feel (about half-way through); Chamber of Horrors and Sell Your Problem (second week); and In Order for Me ... (next to last day).

Topdog/Underdog

In his efforts to integrate the personality, Perls realized the importance of working with polarities. Both from his own struggle with himself and from the language used by patients, he understood that one result of socialization – when the needs of parents are frequently in opposition to the needs of their children

– is internal fragmentation of the child. Parents need children to be obedient and to fulfill the parents' social and emotional desires. Children need to explore themselves and their world, to express themselves (most often in a loud voice and frequently through unacceptable behavior), and to say no as well as yes. The child is utterly dependent on acknowledgment and support from the parents. What a predicament: to silence one's own voice and curb one's needs and desires, or risk losing approval and support. The needs of the powerful quickly become the desires of the weak (Lichtenberg 1990). However, in identifying with demanding, withholding parents, children become alienated from themselves. While they may "look good" on the outside, they deny their "bad" thoughts and deeds on the inside.[3] Statements like "I make myself so angry when I do that," "I keep challenging myself," and "In spite of what I really wanted to do, I did just the opposite" disclose the prevalent conflicts in adults.

Perls called the internal adversaries "Topdog" and "Underdog." Topdog is a metaphor for the aspect of our personality that tells us how to be; it makes demands that we be better, be nicer, work harder; it bullies and threatens us. Perls described Topdog as "not always right but always righteous." Some people identify with Topdog, ready to admit being like a demanding, but socially acceptable, parent. Others identify with Underdog, the metaphor for Topdog's opposite: "Yes, you're right, *but* ..." They may recognize in themselves the recalcitrant child who argues and whines. Underdog's role is to manipulate, exhaust, and sabotage Topdog: "I couldn't possibly do what you ask." "I know you're right. I try but I just can't seem to do it." "But I like chocolate so much." "I can't help it, I'm just like my father." "Tomorrow I'll try to figure out a way to be better." Much energy is dissipated by the two forces' being pitted against one another and the impasse that results. To make the implicit explicit, one experiences the opposite poles and gives voice to both. One actually creates a dialogue between Topdog and Underdog; the internal monologue is made external, bringing increased awareness and mutual recognition. As the adversarial forces encounter one another, energy that had been blocked may be released, and it becomes possible for the person to begin a process of integrating what is often disowned or misused power.

We present the exercise in the following way. Some time before we do the actual exercise, we ask the students to each make a list of five ways they should improve or be different. We give examples: perhaps they think they ought to study more, or visit their parents more often, or not be so negative toward life. Then we tell them to put the lists aside for later use.

Part 1. When the time comes for the actual experience, we ask the students to sit in pairs facing one another. Number One in each pair becomes Topdog; Number Two becomes Underdog. Topdog is to tell Underdog how she or he should "shape up," improve, become a better person. In a mildly ironic tone of voice, we assure Number Ones that they need not hold back, because they should really want to help the Number Twos; indeed, they can feel righteous because of the good that will result. Of course, the students do not know each other very well and are reluctant to be unkind or hurt feelings, but they get into the spirit of the exercise; most enjoy it, and some become aware of essential and meaningful issues. Meanwhile we instruct those in the role of Underdog to respond in any way that seems natural for them. After about five minutes the Number Ones and Twos reverse roles. Then they metaprocess. As we usually do during metaprocessing, we ask questions to help keep them focused on the task and become aware of behavior that may be so automatic it is outside their awareness: What sorts of Topdogs were they? Were they gentle and kind, or very strong and demanding? Could they move into the role easily? If not, what interfered? What kinds of Underdogs were they? Did they argue? Or, did they agree with everything their Topdogs told them even though they knew they would never comply with the Topdogs' demands? Did they defend themselves, or make excuses? Did they notice their voices as they played roles? Did they identify more with the Underdog role or the Topdog role? Do they believe they succeeded in improving their partners? What feelings, if any, surfaced for them? After the metaprocessing in pairs we respond to any comments or questions that arise.

We then describe the procedure for the second part of the exercise and demonstrate it briefly. This requires spontaneity of the demonstrator to set the tone; rather than rehearse in order to remain "safe," we let the dialogue and the process unfold naturally. Then we metaprocess as we expect them to do, both describing the experience.

Part 2. We ask each Number One to carry on the Topdog/ Underdog encounter with herself or himself while Number Two observes and listens but remains silent. Number One takes the list (written earlier) of ways to be better, stands facing an empty chair, and talks to the empty chair *as if* he or she is also in the chair. While standing Number One takes the role of (i.e. becomes) Topdog, looks at the first item on the list, and begins: "I have been telling you for a long time that if you don't speak up, you will never get what you need. It's the same everywhere. Look what's happening in your work and at home." Now Number One sits in the chair, becomes Underdog, and responds to Topdog. "Well, it's not so easy as you might think. Look what happened when I tried to talk to my boss. Nothing. In fact, it's worse now." This person continues to let the dialogue unfold, changing roles as needed, moving from one item on the list to another as time permits. Meanwhile Number Two watches the posture, gestures, and facial expression, listening to the tone of voice as well as the words of Number One.

Part 3. When the allotted time (10–15 minutes) has elapsed, the pairs metaprocess in three steps: 1) Number Ones tell of their experience: what the tactics of Topdog and Underdog were in relation to one another; what, if any, feelings and emotions surfaced; whether they could appreciate their proficiency at distracting, distorting, and giving alibies in the role of Underdog; whether these interchanges were familiar; etc. 2) Number Twos share observations without judgment: whether there was a change in voice, posture, or manner between the two roles;

when there was energy and when the energy seemed blocked; whether Topdog and Underdog seemed to listen to one another; and whether one seemed to prevail or to "win." We stress that the feedback should be descriptive rather than evaluative – thick (i.e. vivid) description of what has been going on. This is not a time for psychologizing, passing judgments, or giving advice. Number Twos also comment on their reactions as Number Ones were working. Then Number Twos have a chance to play out the encounter between the two roles in the same way. 3) After processing the Number Twos' experience, all may reflect on the origins of their "shoulds" and whether they recognize their Topdog and/or Underdog in their relationships with others.

At the conclusion of the exercise they free-write. Then we metaprocess in the large group. We ask some focusing questions – for example, "How were you as Topdog and Underdog in your interaction with yourself compared with how you were in those roles with your partner?" We suggest that instead of trying to do something about their Topdog/Underdog conflicts, they might begin to notice – without judgments – when they carry on the internal dialogues. We ask whether they have questions about introjection. We discuss how typical Underdog qualities may be projected onto others by Topdog teachers or bosses as a way to justify their own harshness in dealing with those in subordinate positions. We raise the topic of contact or lack of contact with aspects of the self.

Rationale. We carry on our internal dialogues repeatedly and so rapidly that we often do not notice how we create and maintain our internal (and eternal) conflicts. In voicing the internal dialogues we make them external and slow down the process. We can experience ourselves more totally, first as Topdog, then replying as Underdog, by paying close attention to what happens in the body, the voice, and the emotions. The conflicts we have with ourselves often mirror those we have with others: parents, partners, children, colleagues, and so on.

At various times during the course we help students become aware when they take an "I should ..." or "You should ..." position, or a rebellious, "dig in the heels" stance. We expect that the students, having consciously identified with both the Topdog and Underdog roles, will be less likely to slip unknowingly into these roles when interviewing or otherwise relating with participators in research studies. They may also become more aware when others take one of the positions in relation to them. Their relationships will be clearer when they own their material and stop projecting parts of their inner conflicts on other people.

Student comments

- "I had been doing the Topdog/Underdog dialogue for some time. Why suddenly did the feelings come?"
- "I know about my inner conflict but I still do it with my husband."
- "When you said don't make judgments, that was a relief. Also, it was important for me when you said, 'stick with thick descriptions.'"
- "When I was free-writing I became aware how much easier it was to be Topdog with myself than with my partner in this exercise."
- "I felt despair while processing. I have been doing this for so long to myself."
- "I heard my mother and father in my voice when I was Topdog."
- "My partner told me I sounded very young when I was Underdog. I felt that way, too; I felt like crying sometimes."
- "I understand why I feel depressed sometimes, now that I hear how I talk to myself."

We bring closure to the experience by addressing any further questions or comments.

See, Imagine, Feel

This exercise uses seeing as a vehicle for awareness of self and of others. It has four main objectives: 1) staying in the here and now; 2) explicitly differentiating between seeing and imagining;

3) becoming aware of one's response to what one sees and imagines and verbalizing it without reasons or explanations; and 4) becoming clear about the boundary and differentiation between self and other. By highlighting the three fundamental aspects of being present – seeing, imagining, and being aware of one's response – and in addition encouraging short, direct communication, we teach the students a formula that can vastly improve contact and communication. Much communication is vague and confusing with no clear separation between those trying to communicate; often the pronoun "I" is replaced by "it," "you," or "there is":

- "There's a lot of anger." Where? *Who* is angry?
- "It's embarrassing when you walk into a room of strangers." *Who* is embarrassed when *who* walks into a room of strangers?
- "You're really annoyed when someone doesn't tell you something," instead of owning one's annoyance, as in, "I was annoyed with you when you didn't tell me."

Another common confusion is between seeing and imagining. On seeing a colleague take a seat across the room, one may think, He doesn't want to sit with me. Problems arise when we respond on the basis of what we imagine: He doesn't want to sit with me. OK, if he doesn't like me, I'm not going to ask him to ... Instead, the person needs to look to see what is going on, or ask, "I'm thinking you don't want to sit with me. Is that so?" People generally see only what they want or choose to see. Smith suggests that "others appear as they do less because of what they're like than as a result of the filters through which they're being examined" (Berg and Smith 1988:123). For the researcher it is important to differentiate between oneself and another and between what one sees and imagines. Staying in the "here and now" facilitates interaction and the collection of data, as does the ability to speak in simple, direct statements.

The See, Imagine, Feel exercise requires students to sit in pairs facing one another. Each, in turn, says three sentences to the other. The first sentence begins with "I see ..." – the person speaking says something he or she sees when looking at the other. The second sentence begins with "I imagine ..." – the speaker says something he or she imagines about the other in the moment. The third sentence refers to the speaker and begins with "I am ..." or "I feel ..." These sentences constitute the entire "conversation." The second person then makes all three statements, and they continue to take turns until we call time.

We voice some cautions before they begin.

Caution #1. Be certain that what you see is truly what you see and not your interpretation of what you see. For example: "I see you are nervous. Wrong! That is what I imagine. What I actually see is you breathe quickly and shallowly, and you fidget with your hands. I *imagine* you are nervous." Seeing requires a person to look at the other, not at the floor, not at the ceiling.

Caution #2. The third sentence is about the speaker's own experience, his or her response. Because the context of the exercise is somewhat neutral, or unemotional, this sentence may seem difficult. We suggest you use your senses. How are you breathing? Where are you tensing your muscles? Is your jaw tight? Your abdomen? Your shoulders? Are you leaning forward toward your partner? Or pressing against the back of your chair? Are you interested, engaged, bored, or amused? There are numerous possible responses.

Caution #3. Avoid saying, "I feel that ..." What follows will probably have little or nothing to do with feelings, but will be a thought: "I feel that this is interesting."

Caution #4. We repeat, the third sentence is about the speaker. Please *do not* begin your sentence with, "I feel that you ..." This will be a statement about your partner, not about yourself; in fact, it will be an imagination.

When we complete our explanation of the exercise, we demonstrate what we have described and metaprocess when we finish, as we ask them to do. We then answer any questions they might have before they begin the exercise.

The students carry on with the assignment for four or five minutes. Then they metaprocess about their experience as they were engaged in the exercise. We ask questions to keep them focused on the task: "Did you notice that, when you imagined something about your partner, it was true for you? For example, 'I see you look at me. I imagine you are nervous, and I am nervous, too.'" This kind of statement may be blatant projection – and it may be true. We also ask them to talk about the question of contact. Did they experience contact with their partners? How do they account for the contact or lack of it? We then ask them to change partners and repeat the exercise. Depending on the time available we take a few minutes for free-writing and then metaprocessing in the large group.

Rationale. As stated in Chapter 4, investigators must be able to pay attention to themselves and others, to listen with interest and empathy, and to respond in the moment. This exercise brings these abilities into focus. There is no task other than to be present in the here and now, in touch with oneself, and responding to another person. There is no "subject matter" to distract students from the task. The formula – see, imagine, and verbalize your response – is simple and easy to remember, can be practiced in almost any situation, and can be used whenever one is interested in improving quality of contact. Like any other skill one practices with a desired outcome in mind (in this case, improved contact), this can be used to manipulate another person's behavior: "I see you look away; I imagine you would like to stop our conversation and I feel very disappointed." This might call for the other to reply, "Oh no, I don't want to stop." By doing the exercise in the classroom context, the complexity of interaction is temporarily simplified, reduced to the bare bones. All qualifications, explanations, blaming, and defending are quelled. Interpretations and imaginations are labeled as such. Metaprocess time after the practice provides what we rarely get in life: immediate feedback from someone who is exploring together with us in a learning context, with little or no emotional involvement and no demands regarding how we should be.

Student comments

- “It was not so easy to go from looking and imagining about the other and coming back to see about myself.”
- “Only now as we talk about it I realize I was projecting. Even during the processing I wasn’t aware of it.”
- “This was very relevant for my research – in interviewing and fieldwork. I think I have used a lot of imagining.”

On one occasion a student remarked she had a hard time with the third sentence. “I felt the same all the time,” she said. I offered to do the exercise with her for a short time. She agreed. I soon commented that she saw my earrings or my clothes but she seemed not to see me. Then she looked and saw my eyes, which previously had seemed too “personal.” She realized that she had not allowed herself any personal contact or response. At that moment she responded.

Chamber of Horrors

This exercise provides an experience in working with “undoing the retroflection,” or doing to others what we do to ourselves. The hostility we chronically direct toward ourselves is seen in Gestalt theory to be misdirected. Beginning early in childhood we react with hostility to those who frustrate us. Our attempts to express impulses that are seen as negative, wrong, or hurtful to others are interrupted. Aggression is punished; consequently we learn to turn in on ourselves the hostility we feel toward others.

The students choose partners and sit in pairs. We ask them to close their eyes for a few moments and think of the ways they “put themselves down.” We ask, “What do you do to your body and say to yourself when you make yourself feel small, wrong, and incapable?” We allow enough time for them to think about the question, and then talk together about it. We next ask them to tell which of their various methods is the most efficient – what, in detail, they do to themselves, with their posture, with their breathing. And finally, what do they say to themselves?

One student in each pair, designated as Number Two, becomes "clay" in the hands of Number One, who becomes the "sculptor." The Number Ones will do explicitly to Number Twos what they do to themselves when they put themselves down. We direct Number Ones to arrange the body positions and facial expressions of Number Twos – actually doing it with their hands – and to tell them how to breathe and what muscles to tense. Then they tell Number Twos the words, in the present tense, they must speak repeatedly. When all Number Ones have completed their work, they walk around the room to look at the other "sculptures" that make up the tragicomedy of the Chamber of Horrors and to listen to their sentences. They see the other sculptures in their tortured positions and hear the messages. Here are some examples of sentences spoken by the sculptures:

- "You'll never manage this. Everyone is better than you."
- "Be aware of your limitations; if you're not it's not good for your nervous system."
- "You haven't got what it takes. Just give up."

Number Ones return to their partners; they sit together to metaprocess. Number Twos begin by relating how they perceived what was done to them when they were the clay and any feelings connected with the experience. Then Number Ones talk about doing the sculpting, doing to someone else what they do to themselves, and their reactions to seeing the other sculptures. Next the roles are reversed: Number Twos sculpt Number Ones and give them their sentences to say. Then Number Twos survey the Chamber of Horrors. Initially they metaprocess what occurred during the exercise, including what effects this experience might have on them as researchers. Then they free-write.

Rationale. Doing to someone else in an explicit and exaggerated way what one does to oneself – usually without awareness – creates a parody of oneself. In the process of doing this one needs to pay close attention to method: What do I do? How do I do it? What do I say to myself? Then one steps back and, from a meta-position, takes a look at this ridiculous, cartoon-like figure and at the other creations around the room. Generally there is plenty of laughter in the room to lighten what has, at the same time, a tragic quality. The students both enjoy the humor and are sobered by their self-cruelty. Having done this exercise, they are less likely to put themselves down without some awareness.

Student comments

- "This gives a lot of energy. You take a metaview and realize you are not alone in this business of putting yourself down."
- "Sharing in the whole group was very funny and very sad. It's very stupid what people do to themselves."
- "I wanted to laugh at myself to avoid getting right into it and seeing myself."
- "I enjoyed the modeling and seeing the others."
- "When I looked at my partner being me, I didn't enjoy it. I got irritated."
- "It was comical and horrible; I had to laugh."
- "It is no wonder I feel depressed sometimes."

Sell Your Problem

If we have a free block of time on the eighth day, we include this experience. The students form groups of four or five. We instruct them to close their eyes and think of a problem they are familiar with, a chronic problem they have been unable to eliminate and are tired of having in their life. Then we tell them to attempt to sell their problem to someone in their group. Each one has a turn to do this. One student tried to sell his father-in-law: "You would always have someone in your life to tell you the right way to do things." Others have tried to sell their habits of procrastination: "You will never have enough time to do anything really well. You will always have an excuse for doing mediocre work." They elaborate, perhaps exaggerate, and generally put much energy into this task.

Rationale. To sell a product we must make it look attractive, valuable, and/or useful. We must promise buyers that this product will do something for them. To sell a problem we make it look desirable; to do that we must reframe what has been experienced as negative to appear positive. This requires shifting one's perspective, taking a point of view different from the habitual one – a metaview. We see this as a valuable practice for those doing social research. They must continually ask themselves, How else could I view this behavior? What other explanation could there be for this phenomenon? If investigators neglect to come up with alternative explanations, they could jeopardize the usefulness of their research. The following is an example from Antonovsky (1994:7) in which he tells of Abram Kardiner's comment upon hearing of the "great frequency of enuresis and other manifestations of disturbance among the kibbutz children" in Israel. "What else could be expected in a society which so totally removes children from their parents?" The sociologist who had gathered the data pointed out "that the children were observed in 1948, when their kibbutz was being shelled day and night, and they were sleeping in shelters." Antonovsky notes that, at that time, Kardiner was one of very few scholars acutely aware of the necessity of keeping in mind a holistic view when considering human development. He writes, "Even Kardiner had his moments of weakness in which he would forget his own broad thinking."

In Order for Me ...

This exercise helps the students to get in touch with ways they might relinquish responsibility and give away their power to those they interview, observe, and otherwise work with in their research. We present this on the next to the last day. Our instructions are as follows:

Students sit in pairs facing one another and choose who will be Number One and Number Two. Number One, in the role of researcher, must begin each sentence with, "In order for *me* to be a good researcher, *you* must ..." We instruct Number Ones to keep their sentences coming fluently, to not think very much but just let them flow even if they do not make sense. Number Two, in the role of respondent, remains quiet. After a couple of minutes we ask them to reverse roles so that Number Two, in the role of researcher, begins each sentence with, "In order for *me* to be a good researcher, *you* must ..."; Number One assumes the role of respondent and remains quiet. When they metaprocess we help them focus their discussion by asking questions: "In which position did you feel more power, more in charge of the relationship? When your partner told you what you 'must' do, were you interested in complying? Did you discover anything about yourself in the role of researcher?"

Rationale. The students "know better" than to need or use their participators for their own emotional needs. However, this exercise reveals the vulnerabilities that could cause them to do just that. When a participator has been made responsible in any way for a researcher, it is because the researcher has projected power onto the participator. It is as if the researcher has said, "I give you the power to make me feel good," or, "I give you the power to make the study exemplary." Of course, this would never be conveyed in an open way; nor would the researcher, or the participator, necessarily be aware of the collusion. The line between the researcher's task and the participator's task needs to be clearly drawn.

In the large group process the students agree that to expect others to do something so they might be effective – whether as teachers, researchers, parents, or whomever – is to deny that they have the capability to be effective themselves.

Student comments. When one student remarked, "I was surprised at some of the things I said," we asked for an example. In response the student recalled having said, "In order for me to be a good researcher you must like me." Others told some of their statements:

- "In order for me to be a good researcher you must be interested in my study."
- "You must think I am clever."
- "You must think this work is as important as I do.... You must give me your full cooperation."

Rationale for group Gestalt exercises

Group Gestalt exercises are adventures; they are also antidotes for the overuse of cognitive approaches to learning. They coax students away from their usual patterns of interpreting, figuring out meanings and what the professor wants, and relying on opinions that have become mental habits. They ensure connections with the affective domain: imagination, feelings, emotions, and fantasy. They encourage process, fluidity, creativity, and improvisation. They are not overly threatening and are often enjoyable methods to engage students in self-inquiry, increasing self-awareness and acceptance of themselves and others. They present opportunities to move into new experiences in which the process unfolds and the outcome is discovered rather than planned.

Everyone in the class participates simultaneously, which increases the safety factor and encourages spontaneity. As they encounter themselves they also encounter one another, whether in pairs, small groups, or the entire class. The shared experiences and metaprocessing (an important aspect of their engagement) increase their knowledge and trust of one another. For each exercise that requires partners we ask them to choose someone they have not worked with before. They come to recognize when they truly appreciate others as being different from themselves and when this is difficult for them to do. They realize there is not one right way to respond; each person engages and responds in her or his own way. They all find something that is both meaningful and idiosyncratic, depending upon how they organize their experience and deal with those issues that may be stressful for them.

Each exercise brings special attention to one of the boundary disturbances, also referred to as interruptions of growth and contact (see p. 44). As noted earlier the tarot card exercise deals specifically with projection. Topdog/Underdog is aimed at introjection, whereas the Chamber of Horrors deals with retroflection. Following the Gestalt principle, we promote bringing parts of the personality that have been denied and powerfully silent into the light of day and giving them a voice, always furthering identification with and responsibility for these parts. The more integrated one is, the clearer all relationships become. The significance of clear relationships for researchers is self-evident. Paradoxically, the power in the projections, introjections, and retroflections diminishes with one's experience and acknowledgment of their occurrence. Less energy is consumed avoiding contact – with these parts of ourselves and with others – and more vitality is available for engagement in life and growth, and for doing research.

Each exercise has its own integrity. Furthermore, at another level, it is part of a larger gestalt directed toward a learning process with specific concepts or themes that seem crucial – specific characteristics of the qualitative research role. Ultimately, each exercise is an increment of the total training to prepare participants to use the qualitative approach effectively. When the components of the course are combined, the constructs of awareness, the now, projection, introjection and retroflection, presence, metaprocessing, reframing, and others are covered, both experientially and cognitively. The course provides opportunities for empathy, self-disclosure, and meaningful interactions with various people in a safe setting. There is much laughter and there are tears. The training can be conceived as a case study of what doctoral students might experience as an important part of their preparation for doing research.

Student comments

The following are remarks made at various times when we asked for reflections on the day:

- "The hours go so fast. I'm very involved; it's all so meaningful. It does me good and I have a good time."
- "I'd like more discussion on specific topics from the reading, about taking more responsibility. I look forward to tomorrow."
- "Today was a very good day. I've learned new things, had new experiences. The day brought many surprises about how others see me in contrast to how I see myself."
- "I'm tired. I worked hard this morning. I'm proud of the work I did – that I did it in front of others. I feel more safe and connected with you all."
- "It's especially good for me how you keep connecting all this to research. It helps also to keep connections with the book."
- "It is good when you connect what happens in the working chair [process awareness work] with the researcher – how we are all motivated by unknown influences, both as researchers and subjects."
- "I have written about learning from experience, but now I realize I didn't really know what experience is. It was all in my head."
- "I saw something I never saw before. I saw the group become a group."
- "In the chair work you do something so helpful when you just stay with the process and make things explicit, like when you said, 'At this point you interrupt your process.' Such comments are very illustrative."
- "One of the ways the book was important for me was that Berg and Smith and the other authors put words to things so I could make connections: Oh, that's what it is; that's what is happening. And when you used the word 'resent,' it was a magic word for me. Since then I've been able to work my way through stories in my life and help finish them up. So many pieces fell into place with that word."[4]

Summary

Students vary in their readiness to engage in this course, where self-awareness is a major focus. As instructors, ever conscious of the students' reactions, we encourage them to voice all questions, disagreements, and doubts that arise. The posture of the instructors is a key to the success of such a course: a non-judgmental attitude must prevail, along with a willingness to engage with the students, serious concern and respect for each student, and a quick sense of humor.

The cognitive dimensions of the course – reflection and discussion of the text (Berg and Smith 1988) and theoretical interpretations of the Gestalt approach – are enhanced by process awareness work and training.[5] These activities provide non-cognitive comprehension and create a climate of self-disclosure, acceptance, and contact, all important to the role of researcher.

I have described the affective content in some detail, not as "cookbook instructions" to be followed step by step by others who may embark on similar courses, but simply to put forth what has constituted this particular course for these students. It may be used by others as an example of what is possible and altered according to their experiences and creativity as well as the needs and readiness of their students.

A word of caution: The integrity of the instructor(s) is of utmost importance. They should metaprocess the class during and at the end of each day to provide opportunities for the instructors and students to keep the lines of communication open. In other words, the class should involve instructors and students in an ongoing exploration by way of frequent, systematic opportunities for metaprocessing. We continually modify our course according to what emerges in metaprocessing.

Chapter 6

Process Awareness Training

Attention equals life.
Edwin Denby[1]

A major focus of the course is process awareness training. This consists of laboratory experience in the arts of seeing, hearing, personal engagement, eliciting meaningful information, and giving descriptive feedback. We devote the major part of several afternoons to the development and practice of these interactive skills – skills essential for establishing presence and bringing the participator (also referred to as "the client") and the researcher into the "now experience."

Although we seem to dismantle communication to examine some of its parts, thereby going against a holistic view, the objective at each stage is to increase the students' awareness of mutual influence, connection, and awareness of self and other, to better appreciate the complexity of even the simplest, most contrived interaction. Process awareness training is practice in paying attention to oneself and others, listening with interest and empathy, and responding in the moment. The researcher, while "in the role of" interviewer, does not actually play a role but, instead, is available for human-to-human connection.

The structure of the training was developed over many years as we conducted training for professionals in various fields, primarily psychotherapy, education, and medicine. What I present here is basic training in which, as we focus on essential threads from the rich tapestry of human communication, we always deal with the larger issue of relating. We make clear from the beginning that we are teaching the students not to become therapists, but to acquire and use interactional skills to improve their research.

Seeing and hearing

We begin process awareness training by introducing the basic skills of seeing and hearing. In presenting each step we follow a pattern in which we

- give a brief background;
- explain the structure of the mini-lab;
- instruct the students to form groups of three for the roles of investigator, participator, and observer;
- describe the roles and tasks;
- demonstrate the roles of investigator and participator; and
- begin the exercise.

When all three members of each group have assumed the three roles and completed the final metaprocessing (which, as explained in Chapter 4 and elaborated on in Chapter 5, is communication about people's experiences, their relationships, and the rules and patterns of those relationships), we ask them to take a few more minutes to talk about the significance of the training for their research. Then they free-write about the experience and its connections with their research, whether proposed or currently ongoing.

In this first exercise we focus exclusively on seeing what is – not what one wants to see, not what one thinks one ought to see, not what one imagines.[2] We acknowledge to the class that the interaction in the mini-lab might seem bizarre, even silly, and suggest they put aside their habits of interaction and thought and move into this new experience. We ask them to form groups of three and to decide who in each group will be Number One, Two, and Three. Every person will, in rotation, be in each of the roles: investigator, participator, and observer. To begin, Number One as investigator and Number Two as participator sit facing one another; Number Three as observer sits to the side in order to watch both of them.

Each has a specific task. Number One observes the body posture, movements, and facial expressions of Number Two and copies these to give non-verbal feedback. This is "mirroring." At certain moments Number Ones also get in touch with their experiences during the process, that is, how they are as they take on the postures, movements, and facial expressions of the participators – quickly shuttling between self and other. We remind them to notice the entire body, not to become fixed on the eyes of the participators nor distracted by the content of the words, for this interferes with seeing in the beginning. (Later, with practice, they can do both.)

Number Twos, as participators, report whatever emerges for them in the moment. They may be distracted by the feedback they get from their investigators; if this is happening, they may say so. They may have thoughts from the morning session, or an awareness of the temperature and sounds in the room. All these are appropriate as "now" statements. Number Twos are not to make up fiction or tease their investigators by intentionally and artificially making faces, taking strange and difficult postures, or gesturing; rather, they should simply be natural and tell their experience, even if what they feel is emptiness and they do not know what to say. In that case they simply report, "I don't know what to say."

Number Three observes both One and Two in preparation for giving verbal, non-judgmental feedback to the investigator at the end of the session. The observer, being outside the interaction in a metaposition, is able to witness the interaction, which becomes more complex as we move through the training sequence.

After giving instructions, George and I demonstrate the roles. We decide who will be Number One and who will be Number Two. How we engage sets an example for the students. We do not play act, but rather carry out the roles spontaneously and authentically. After our interaction as investigator and participator, we metaprocess and ask for descriptive comments from the group. We remind them to practice non-judgmental, non-evaluative feedback and to notice the messages in their voices.[3] Then they begin their mini-sessions of three minutes. Sometimes we need to prompt the participators to start talking. As they engage we wander from group to group, noticing, encouraging, sometimes reminding a Number Two to talk or a Number One to mirror a smile or gesture of the participator.

When the three minutes are over, they metaprocess the exercise in a special way. First the investigators tell their experience – whether it was easy or difficult; whether they felt foolish; and what they were aware of as they mirrored the participators, including the physical experience as they assumed the postures, gestures, and facial expressions of the participators. We remind them to speak about themselves, not the participators, in this metaprocess session. Next the observers tell the investigators in a descriptive, non-judgmental way – being aware of their voices and whether they carry messages of judgment – what they noticed from their metapositions (e.g., "Your head kept moving as if you were agreeing or encouraging your participator, but your participator's head was not moving." "You moved your feet in the same way as the participator." "Did you see how your participator was moving her mouth? You didn't copy that."). Finally the participators tell their experience, especially how they responded to the investigators during the session.

After the metaprocessing, which lasts about five minutes, we ask them to rotate: Number Twos become investigators; Threes, participators; and Ones, observers. We continue in this fashion until each has been in all three roles and they have metaprocessed for the final time. In the large group we respond to any questions and comments they may have. We suggest they pay attention to how they see outside class and whether they interfere with their seeing by, for example, imagining, thinking about the words of the other, getting caught up in the story being told, formulating a response, or simply not looking.

Student comments

- "I suddenly felt immobilized. I saw the participator move but it took me a while before I could do both, see it and mirror it."
- "I thought it was mean to copy the participator. It reminded me of when I was a child. I didn't want to do it."
- "I wanted to see everything, even the eye blinks. I felt like I was in the other body and could anticipate her moves."
- "As participator I was surprised to see how active I was."
- "I felt like my investigator was exaggerating my movements and I didn't like to see that."
- "When I was participator I saw from my 'mirror' how much I smile – in fact, all the time."

The students form new groups of three for the next step in the training: focusing exclusively on hearing the voice. Listening, like seeing, is a subjective activity. The listener participates by making meaning of the words, the voice, and the stresses on particular words, by being sensitive to the emotional message behind the words and the choice of words spoken. For example, some words have a formality, others are polite without being formal, some are intimate, and so on.

The students determine who will be Number One, Two, and Three in their new groups. Ones and Twos sit opposite one another. We instruct each investigator to give feedback on the voice of his or her participator by voicing as accurately as possible, without words, only the sound of the voice, including volume, pitch, melody, rhythm, and whether successive sounds are disconnected (staccato) or connected (legato). The investigators do this by using syllables such as "da-da-da" so that the mouth is open. They are also to shuttle their awareness back to themselves from time to time and experience their throats and chests as they imitate their participators' voices.

Before they begin we demonstrate what they will do in their groups. We also give them the opportunity to imitate the sounds of our speech to practice and loosen up a bit. Some students are shy and hesitant to let a sound come out; having the entire class do it at the same time diminishes their reluctance. And they always seem to enjoy that we leave ourselves open to their "da-da-da" imitations of us. We time the sessions, again allowing three minutes, and call for metaprocessing after each one. After all have been in each role, we allow time for free-writing.

We then metaprocess in the large group. We discuss the importance of the voice as a source of information. In many cases the quality of the voice tells more about the person than any other source of data; we hear the level of energy, tenderness, fear, anger, sadness, hostility, authenticity, and artificiality. We hear when the voice is so low that one must strain to hear it, or so loud that it is overwhelming; when it seems connected to the beingness of the person, or seems to float disconnected on the air; and when the quality of the voice does not fit the content. We speak also of reactions one has to the manner of speaking, for example, impatience with a person who hems and haws and never finishes a thought, or breathlessless with one who rushes through sentences. We suggest they listen more intentionally to voices while outside class and notice how they respond to what they hear.

This first day of the mini-labs evokes excitement among the students as well as exhaustion. These exercises require close attention and elicit a high level of engagement. We sometimes remind those who are in the role of investigator to relax and breathe.

Combining seeing and hearing

On the second day of mini-lab training, we combine mirroring and voice feedback. We use the same procedure for the process awareness training as above: groups of three; designations of investigator, participator, and observer; and three-minute sessions each followed by five minutes of metaprocessing, done in the same sequence of investigator, observer, and then participator. Because there is so much to see and hear, we introduce the idea of *selecting* from the mass of material the participator puts forth. (This is a problem frequently faced by researchers in the field.) The investigators mirror and give voice feedback on what emerges most clearly, or forcefully, into figure for them. We suggest they notice especially changes in the voice, movements, expressions, and gestures that seem incongruent (e.g., smiling while one says she or he feels sad). The observers now have more to pay attention to than earlier. Each is to give non-judgmental feedback to the investigator concerning what the investigator reflected back to the participator and those things that were in figure for the observer but were ignored by the investigator. It is to be expected that two people will differ in what comes into figure and in their feedback. What the investigators select depends on their subjective framework and is neither right nor wrong.

The training evokes from the students many reactions that the metaprocessing may leave unresolved. Free-writing allows time to note some of the ideas, emotions, questions, and unfinished situations. In addition we use the large group to deal with this material. Many questions pertinent to the role of the researcher are asked and discussed. Students comment and ask about how the awareness training relates to their work and, specifically to them as researchers, interviewers, and observers in the field:

- "How can I separate my material from that of the person I am interviewing?"
- "How will I know if I am manipulating people for answers I feel comfortable with?"
- "Why is it that sometimes I can identify with others and sometimes not?"
- "I have a problem with messing around with someone's life. I feel embarrassed when I give back an interview full of my descriptions."
- "I worry about that, too. When I worked as a counselor I had to write reports of what the participator and I did together. I wondered what would be their reaction to what I had written."
- "When does my personal experience help me to understand and when does it distort?

We offer more self-awareness questions they can ask themselves:

- What is my bias in this situation?
- Do I experience any threat?
- Do I have a desired outcome?
- Am I able to reframe the situation? Am I able to take one or more opposing views?
- Where does my safety lie?

Stating the obvious

The third training lab focuses on "stating the obvious." We define the obvious as that which one sees and hears. The investigators now use words to give feedback to the participators. We remind them that, at this point, they are not to concern themselves with being "helpful" to the participators; they are to focus on seeing and hearing and, rather than mirroring and giving voice feedback, they are to use words to tell what they observe and hear. Instead of leaning forward when the participator leans forward, the investigator should simply say, "You lean forward." Instead of saying, "da-da-DA-da," he or she says, "When you say, 'I don't like that,' you accentuate the word 'like.'" We repeat to the investigators that they should not try to comment on everything the participators do and say, but select what emerges into figure for them. This problem of selection from all the available data is a constant one in qualitative research. We recommend they sit back, breathe, let the data come to them, and not try to be perfect. We caution them not to get caught up in the stories and to remember to listen to the voices.

The observers' feedback to the investigators should always be descriptive and non-judgmental. They should now mention if and when the investigators interpreted ("You don't like what is going on," instead of "You lean back in your chair"), imagined ("You don't like me when I do this"), or gave reasons ("You hesitate because you are unsure of yourself"), and in other ways moved beyond stating simply what was obvious. Now that the investigators use their speaking voices in relation to the participators, the observers give feedback to them about their voices (e.g., difficult to hear, lifeless, lively) and whether there were judgments or evaluations conveyed through their intonation, choice of words, or emphasis (e.g., "You *never* stop moving your hands").

Student comments

- "The story was in figure, the rest was ground. I didn't see much."
- "It was easier to be accurate when we mirrored and used voice feedback. It is hard to just describe. I was interpreting everything I noticed."
- "I think my descriptions were too long. The participator would say one sentence and I would say three."

Personal response

This fourth day of the training is difficult and extremely important: The investigators become a part of the interaction in that they become *responsive* and thereby more present for the participators. They do this by voicing how they themselves react to the participators and the interaction in which they now both engage. It is impossible to say what the investigators *should* respond to. Their responses depend on how they react to their participators and to what they react. Perhaps the participator's voice is hypnotic and the investigator begins to get drowsy; maybe the participator is amusing and the investigator laughs. The investigators verbalize their responses. There are no "right" responses, just infinite possibilities. The investigator might be interested, or find it hard to stay in touch with the participator; he or she might be amused, or nervous, or bored. We repeat the purpose of the training: to sharpen the students' interpersonal skills, not to become therapists.

Paradoxically it is at this stage, as the investigator becomes more involved as a partner in the interaction, that the participator's engagement is heightened. The ultimate purpose of the investigator's response is to serve the participator – to facilitate his or her process, add depth to the interaction, and increase contact. During this practice, however, we advise the investigators not to worry about whether their personal responses are facilitative. It is sufficient for them to see and hear the participators, shift their awareness to themselves from time to time, and voice their responses.

We alter the structure of the mini-sessions at this point. We advise the students of the total time they have for the mini-lab work (perhaps fifty minutes or an hour); each member of the group should take a turn in each role, and a metaprocessing period of five minutes should follow each session. They are responsible for keeping time and calling for the end of sessions, which may last as long as fifteen minutes. The observers now have an added task: They are to interrupt the sessions, just once, after three or four minutes and give descriptive feedback to the investigators. This is not a metaprocess period, but rather one minute or less of comments about the investigator's voice and posture, perhaps along with a reminder that he or she has not used personal feedback, or stated what was obvious. We have found these short interruptions extremely valuable. The investigator and participator then continue the session until one wants to end it, or the observer says their time has run out and it is time to metaprocess.

We caution the students regarding the personal feedback of the investigators, reminding them how personal comments may be manipulative. Participators detect or imagine expectations of interviewers and are influenced by them. Rice (1929) showed the effects of these perceived expectations; through subtle cues, facial expressions, and the way questions are formulated and followed up, one may prompt a participator to reply in a particular vein. Researchers must realize the danger of misusing what Berg and Smith refer to as "the complex emotional and intellectual forces that influence the conduct of our inquiry" (1988:11).

Following our usual procedure, after giving instructions we demonstrate the addition of personal response to the repertoire of stating the obvious. This demonstration is an eye-opener for many students because the interaction becomes much more "normal." Since we are a married couple, some are nervous lest we become embroiled in a personal argument provoked by one of our responses. They observe, however, that regardless whether George or I take the role of investigator, that person states the obvious, gives personal response, and does not get distracted by the content of the other's words. The investigator does not become provoked, nor defend, nor explain. (This is not to suggest, however, that when we demonstrate the participator intentionally tries to provoke the investigator.) We metaprocess our interaction and invite the students to be our observers. They comment on our interaction, on how we allow the process to carry us rather than try to control or direct it, yet stay within the limits of the exercise.

With the addition of personal response we notice a higher level of engagement in the small groups. An increase in the students' trust and understanding of one another is evident as they move into this much more human relationship for extended sessions. Many students, however, have difficulty and may feel awkward during this first attempt at sharing their personal responses.

Student comments

- "As investigator I gave positive response only. I wouldn't say anything that might be experienced by the participator as negative."
- "As observer it was difficult to observe the investigator and the participator. I couldn't keep both in figure at the same time."
- "When it comes to personal response I'm not very good. It was easier to see what was going on from the position of observer."
- "I got very little personal response from my investigator. I felt very uneasy. He wasn't engaged."
- "My investigator wasn't giving much personal response in *words*, but in every other way."
- "All of this makes me very aware of my interpretations of my students when I teach."

- "This was fascinating, being on two levels – separate from the participator when I stated the obvious and connected when I gave personal response."

Because of the unfamiliarity of giving personal response, many students feel they need more time to "catch on." For that reason we take extra time to repeat the mini-lab practice of combining personal response with stating the obvious. On this second day of practice, they are relieved to find that they overcome some of their difficulties and even enjoy the process as they see the potential of their newly acquired skills.

Basic questions

Researchers are accustomed to asking questions, but often the questions they ask do not elicit the information they wish. The question "why?" is frequently used, almost automatically, when one does not know what else to do. We strongly advise investigators to refrain from using "why?" It is not a productive question when dealing with human behavior and interaction. Let us consider the following interchange:

A: I'll tell her it's OK, that I really don't mind.
B: [perhaps wanting to know A's motivations or desired outcome] "Why say that?"
A: [Hearing an accusation that she is about to do the wrong thing and is not aware of her motivations or intentions, she becomes defensive and gives one or more excuses, explanations, or rationalizations for her feelings and intended behavior.] Well, you know, she really needs someone to be friendly to her.

A's explanation may be partially true; however, there is rarely only one cause behind someone's behavior. So A chooses a reason that makes her seem like a good, decent person, in her own eyes as well as in B's. However, B will probably be dissatisfied with A's answer and respond with, "But that's crazy," or, "That won't do any good," or, "That doesn't make any sense at all."

At an early age children begin to plague their caretakers with "why?" At age 3, one of our granddaughters started saying, with great urgency, "I really want to know *why*. *Just tell me why!*" Used by adults the question closes doors to further communication more often than it opens them.

The four questions we consider most contactful and productive are

- What are you doing now?
- What are you feeling now?
- What do you experience now (in your body)?
- What is happening now?

These "now" questions help bring interactions into the here and now and make the people more present and responsive. Even when the content focus is on the past or future, it is here in the present that our encounters takes place; we respond *now* to the thoughts and feelings evoked by the past or future.

On the sixth day we expand the process awareness repertoire of stating the obvious – that is, what the students see and hear, and giving personal response – by adding the basic questions. Some find the questions a relief; asking questions is something they are comfortable doing. As in previous days the students organize their time in the small groups so that each has a turn in all three roles. The observer interrupts to give a minute or less of descriptive feedback to the investigator after three or four minutes have elapsed; then the investigator and participator continue to the end of the session. Either of them may suggest they close the session, or it may be stopped by the observer, who is also the timekeeper.

After giving instructions we demonstrate adding the basic questions to stating the obvious and giving personal feedback.

Worst interviewer

Toward the end of the course we introduce the "worst interviewer" experience. The students are again in groups of three, in the roles of investigator, participator, and observer. The investigator has two instructions: 1) Be the worst interviewer you can possibly be, and 2) enjoy it. We instruct each participator to respond authentically, not to be a "worst participator." The observer is silent until the completion of the mini-session, for which we allow about four minutes. The groups metaprocess after each mini-session, in the usual sequence. The investigators talk first about their experience, including what sort of worst interviewers they were and whether they could enjoy it; they also reflect on the qualities and characteristics they displayed. We ask them to consider whether they discovered something they could use from their worst interviewers in doing serious work. Then observers give feedback to the investigators, telling what they saw and heard from their metapositions. Finally the participators relate how they experienced the encounters.

Rationale

We have noticed how students in the role of interviewer often "try to do it right." In their trying they tend to be exceedingly earnest and careful, which negates all their inherent vitality as well as authentic contact. This exercise provides an opportunity for interviewers to relax and enjoy the experience; the instructions tend to release them from feeling they must "do it right." Their creativity and humor surface, as do unacknowledged aspects of their personalities – polarities of their dominant qualities. In this paradoxical situation, to be bad is to be good. Very considerate people, who normally want to take responsibility for their interviewees by being very "helpful," become neglectful, inattentive, and thus more likely to force participators to mobilize their own energy. The students in the role of participator who face these worst interviewers can also make important discoveries. Students have reported that the "best interviewing of the week" occurs during this exercise.

Student comments

- "I found it very difficult to be 'worst interviewer.' [Another student interjects, "She was terrific!"] Now I think of many things I *could* have done."
- "I loved it! I felt myself very creative."
- "When I was my worst interviewer it was right out of my repertoire. I probably do some of this in my life and don't realize it."
- "As participator I learned the most important thing for me. I really saw how I make myself weak. It was good for me to see it so clearly."
- "I got some awareness of doing double-bind communication.[4] I wasn't aware of it until my participator gave me this feedback."
- "When she was worst interviewer I began to feel insecure; it didn't come into my mind that it was in her – that the 'authority' had the fault. I felt *I* had to do something."

Subject interview

On the next to final day, when the students have practiced their entire repertoire of skills and the "worst interviewer," we give them an opportunity to use their skills in a different format.

We have them sit in pairs; one becomes interviewer and the other, participator. The subject of the interview is the participator's being a member of the class and her or his reactions to the course, which is almost over. The interviewer is to use the skills we have presented during the course. The session may last twenty minutes or more and is followed by metaprocessing. During the metaprocessing the students are asked to consider

- whether they had stayed in the present situation or talked of past times and places;
- whether the interviewer had been present;
- the quality of contact between the two;

- whether the interviewer had been able to collect thick description;
- whether gender had in any way been an issue (especially in a female–male couple); and
- whatever else seems pertinent.

At the completion of the metaprocess session, the roles are reversed and the procedure repeated.

Student comments

- "As participator I liked the chance to look back over the weeks. Also, I was aware of not going automatically into old patterns but staying aware of the task."
- "Maybe we should have done this sooner and then gone out and interviewed a person outside of class. It would be good to go with an observer."
- "I like this work. It is good to create questions ourselves, but not start too early with that. Then we go into content too completely. First we need the practice of the basic skills so we don't just get caught in the content."
- "I noticed when I gave myself time to give personal response and state the obvious that I didn't have to ask so many questions. More information came."
- "It was like a summary exercise – a culminating experience."

Summary

The use of mini-labs for experiential learning is based on a confluent education approach. Learning in this context is a more total experience in that the lessons are structured to include both the cognitive and affective domains. The students engage *intra*personally, increasing self-awareness through their emotions, introspection, and reflection. In the *inter*personal sphere they increase their capacity to share and be in relationship, to experience multiple meanings, mutuality, and empathy. In the non-judgmental atmosphere they come to recognize the rules they bring to the training situations and their own tendency to be self-referential. They frequently come to appreciate differences between themselves and others. During our ten days together, students report with surprise that they begin to react in new ways to people in their lives outside class and that the interactions take productive turns they had never expected.

Self-awareness and the abilities to listen, be in contact, and maintain a separation from the system under study *while* fostering participatory consciousness are highly praised in qualitative methodology literature. An emphasis on those abilities permeates our every interaction with the students. Through the training in moment-by-moment awareness, the students begin to realize when they lose their capacity to listen and when their emotions are tapped in ways that cause knee-jerk reactions, which hinder their ability to pay attention and blind them to the available options.

The express objectives of the process awareness training – honing the skills of seeing, hearing, and giving personal response – are augmented by the less obvious but equally important learning for becoming a researcher. The students learn to move rapidly from one focus to another; each becomes, in quick succession, an investigator, a participator, and an observer. They work with a variety of people, since we change the makeup of the small groups daily. They discover the importance of finishing one situation before they can be present in the next. They come in contact repeatedly with their own material, that is, their needs and fears, their psychological defenses in relation to others in simplified, formularized, safe interchanges of short duration. They hear their partners' reactions and responses to the same interaction, all different, all idiosyncratic – three different experiences, all "absolutely right," as in the story in Chapter 5. Several times each day they experience appreciation of differences, self-disclosure, descriptive orientation, flexibility, humor, acceptance of their and others' emotions, reframing, and – obviously yet notably – moving into new experience. All of these possibilities become actualities, not for all of the students all of the time, but for all of the students some of the time.

The step-by-step description that constitutes most of this chapter is the bare bones of the process awareness training. It is like a transcription of a conversation, devoid of all nuances and non-verbal exchanges. The whole of any entity cannot be broken up without losing its character. So it is with this course. The interaction among the students in the mini-labs, during both the investigator–participator encounters and the metaprocessing sessions, develops quickly into a rich source of learning. That which on paper is a stark, formularized exchange becomes transformed by the students into an intense and engaging awareness-building interaction.

Chapter 7

Thick Data from Students

This shaking keeps me steady. I should know.
What falls away is always. And is near.
I wake to sleep, and take my waking slow.
I learn by going where I have to go.
Theodore Roethke[1]

There is a Russian proverb: If you throw enough mud on the wall, some will stick. We present many ideas and experiences in these two weeks. We cannot know with certainty what "sticks," what the students take in, how they interpret what we say, what makes an impression, or how they integrate the classroom experiences and transfer them to their professional and personal lives. Although we get a general idea of how people are responding to the course from the many metaprocessing sessions and from sitting in with small groups during journal reading and process awareness training, we realize that students filter their experiences through the lens of their own history.

The crucial question for us is whether the course is found to be useful and, if so, how the participants actually apply what they learn. An ideal way to gather the most reliable data regarding the effectiveness of the course would be to observe how each participant functions in field situations. However, this would not be practical. The next-best approach is to ask former students how the course affected them. To solicit some answers to this question, we sent letters in September 1993 to those people who had attended the courses given in September 1992 and in

May 1993 (see Appendix C). Eighty percent of the students responded.

We found certain aspects of their responses highly illuminating:

- their willingness to write the reports, which were in no way required;
- their exact descriptions of how they were using what they had learned about using themselves as instruments in doing field research; and
- their metaperspective reports about what was happening as they prepared for their activities, carried them out, and reflected on what they were doing and had done.

Their letters indicate that the acquisition and application of this learning occur on two levels. The first is the *skill level.* This refers to specific awareness skills such as seeing, hearing, focusing on the "now," and other abilities emphasized in the process awareness training to foster participatory consciousness. These skills, individually and in combination, along with theoretical issues, provide a base for the second level. This might be called a *structural* or *organizational level,* where the approach as a whole functions as an operational gestalt, a fresh perspective from which to view the world and organize one's professional experience. This includes

- taking a metaposition and reframing;
- perceiving relationships between entities;
- the importance of the ground giving meaning to what is in figure;
- functional rather than cause–effect relationships;
- appreciation of the present as the eventful moment; and
- process orientation.

The integration of the deliberate use of the metaperspective is essential at this level, as are concepts such as personal responsibility and the boundary disturbances – projection, introjection, retroflection, and confluence.

In the four courses conducted before this writing, two students dropped out. One was a woman, Anna (not her real name), who left the first course after six days. On her last day she had been deeply upset by the process awareness work when a student talked about the death, some years before, of a person very dear to her. Anna was overcome with grief over her own loss, the previous year, of someone close. She decided not to return.

The other person was a man who did not return after the seventh day of our second course. In an unsolicited letter we received two days later, he took responsibility for his actions and expressed regret over having left the course; he simply stated he needed to do this. He concluded by expressing his "respect" for and "gratitude" toward the group and us for a "voyage into the wilderness of my emotions and ambiguity," and added that the course "has been very valuable to me, personally and professionally." In both instances the remaining students felt, as we did, the empty place left in the group and expressed concern and regret that in some way we had not been able to provide what they needed.

With the students' permission I include portions of their replies to my letter asking whether the course had proven helpful and, if so, how. The researcher who wrote this first letter emphasizes how she has built into her work a constant shuttling of focus from the *content* of the spoken communication to the *relationship* between herself and the interviewee. To do this requires taking a metaposition.

> In our research work my colleague and I are following 13 students through the four-year study of teacher training, having personal interviews with them once a year.... Our main question is to find out about how the study prepares the students for their future work as teachers. In the interview situation the importance of the self as a researcher is evident. I am writing a personal journal after each interview, where I'm focusing on some basic Gestalt concepts to structure the writing of the journal. The concepts are
>
> – here and now
> – awareness of self

- awareness of the other
- my own shoulds
- my own judgmental attitudes

> In the journal I try to be as frank as possible. What do I think of the interview situation? Boring? Interesting? Nice? Did I like myself in the situation? Did I like the other one? What about dislikes? How was the general atmosphere and how did it develop? How was the other dressed? How did he or she talk? How about my dress and language? Did I ask leading questions? Was I open and confident in the situation? Did I have a hidden agenda? Did I try to control the situation in some way? Did the other try to manipulate me some way or other? In short I do a meta of the interview situation with the five concepts as leading starts. During the interview itself I constantly do a meta for myself along the same concepts or openly to foster the dialogue between the two of us. We have made two rounds of interviews up till now, the fall of '92 and '93.

This letter describes the investigator's use of skills from the process awareness training in her interviews: a focus on the here and now, awareness of self and other, and sharing her personal response and metaview to facilitate the process. In addition, this researcher has apparently incorporated the second level – the holistic perspective – in working from the premise that the interview is a mutually interactive event, that the interviewer and the interviewee and the context ("here and now") form a system. She realizes that simply by being who she is, she casts her influence on the research situation, the interviewee, and the data she collects. At the same time she notes how she is influenced by the person she interviews. Some of the questions she keeps in mind help her to be more aware of what might otherwise be unnoticed aspects of the ground from which she picks up impressions. For example, she notes "general atmosphere," "dislikes," and her own attitudes and behaviors. The high level of the interviewer's awareness of self and the other assures that numerous non-verbal cues, as well as the more obvious behaviors, are noticed.

The next letter has quite a different focus – subjectivity, or the idiosyncratic nature of qualitative data gathering in contrast to questionnaire-type objectivity:

> What is integrated in me from the course is 1) an awareness of how different people pick up different things in the same situation, and the degree of influence that [has on] the data collected. That makes me underline still more that open questions in interviews and observations have a personal color, and that this color is explicit for the researcher, and has to be made explicit for the reader of the document; and 2) the understanding that methodological questions also consist of working with oneself as a person.

The writer of this letter points out the self-referential nature of all encounters – how our perceptions are colored by the prism through which we perceive our world. The investigator, as the instrument through which data pass, is required to do the personal work necessary to 1) recognize her or his prejudices, biases, and blind spots; 2) recognize the consequent incompleteness and/or contamination of data; and 3) render the findings of an investigation more credible by whatever means possible. For example, Gregory Bateson, in his book on Iatmul culture, says, "I myself have so little appreciation of time that I omitted almost entirely to enquire into ... Iatmul calendric beliefs" (Schoen 1994:170). Supporting Bateson's comment, Schoen adds, "Characteristic concepts of [Bateson's] developed thought – pattern, double bind, ecology, mind as a category of living – also leave time out" (p. 170).

During all group processes (e.g. journal reading, process awareness training), the participants perceive the diverse meanings attributed to given situations by others. Perls (1969) called making meaning "a creative process." The appreciation of how each person is involved in creating her or his own special, original experience reinforces the necessity of explicitly including the investigator's experience in all stages of a research study.

The following letter describes how a researcher's approach to methodology and his conceptual framework for examining his data were influenced by the course:

> As you may know I am, in my doctoral program, working with the teacher as an innovator in special education. The teacher usually works with children with different kinds of disabilities, and it is important that he in his work may increase his competence in giving the child the most

suitable education. The teacher's reflection on his own teaching (know-how), the support from colleagues and administration, awareness of self and others and the teacher's cooperation with different kinds of professionals will probably be of importance as mechanisms for increasing the teacher's competence.

From my earlier practice as an educational psychologist and from a pilot study one year ago, I have noticed from my interviews with teachers the importance they pay to their relationship with different kinds of colleagues. They often have very clear opinions about which people they can work together with and which mechanisms may facilitate or suppress an innovative program. Some teachers seem very quickly to adjust to colleagues' understanding of a problem situation, while others only seem to support proposals within their own frame of reference. When focusing on change, it is interesting to look at both problem formation and problem solution. Keywords from practical situations may be: more of the same solutions, simplifications, paradoxes, second order changes, reframing, etc.

Since the time I attended the course in Oslo this year, I have used much of my study time in reading literature from the field of qualitative methodology as a preparation for my own qualitative interviewing and field work (Berg and Smith, Patton 1990). After attending the course, I have worked both with the theoretical and methodological parts of my dissertation. I went back to the theoretical parts with increased attention to mechanisms related to innovation, and have made a new version of the teacher as an innovator in special education, and a new chapter on a theory of reflection.... My own reflection on the [methodological parts of my research] has been concentrated on how to bring about those interviews and how to conduct my field studies. My intention before I attended the course was to talk with the teachers in a non-structured way, using focus areas as a guide. I would try to be a good listener and give supportive personal comments to the teacher's elaborations and reflections on his work. Since the course I have been more aware of my own possibilities in creating interviews with a greater focus on keeping in touch with the teacher's own inner thoughts when working with special education. And for me as an interviewer it will be important to pay more attention to my own seeing, hearing, personal response and basic questions in the situation. This way of making contact and communication will probably create an atmosphere during the interviews [somewhat] different from my first plans. The trust of the process and the moment to moment experiences will probably give some other and some additional information to my study.... I will use free writing and a personal journal immediately after interviews and field observation.... I have also tried to pay more attention in which ways my own personal material may affect my clinical work and my work as a researcher.... I

> have found the course in Oslo very important to me both professionally and personally.

The alterations this man makes in the scope of his theoretical foundation show perceptual and organizational shifts. His "key-words" for describing "practical situations," for example, deal with *relationship* rather than *individual behavior;* he speaks of a paradoxical perspective and reframing, both indications of taking a metaview and dealing with higher levels of complexity. He envisions incorporating process awareness skills in his interviewing with the expectation of being more present and responsive, of involving himself more personally in the interview situation. His expressions "trust of the process" and "moment to moment experiences" are cues of second-level learning and a potential change, not only in structuring his thinking, but also in actual procedures. This researcher's application of theory and methodology speaks directly to the purpose of the course – to improve the quality of research.

What is taken from the course and how it is applied vary according to the students' readiness and needs. We frequently asked the following question during metaprocessing sessions after various classroom experiences: "How might these ideas or this experience be significant for you in your work?" Some applications are unanticipated:

> I do attack written texts in a different way. Whether that was brought about by the course, or whether it would have come anyway, I don't know. I'm much more relaxed when I read now, in the sense that I allow certain themes, paragraphs, statements or propositions to capture my interests instead of striving to grasp the whole content.... I now teach this to my students here; they frequently are worried about learning the "expected thing" when reading. I combine this (very fruitfully, I think) with the hermeneutic concepts of pre-understanding and "horizon" of understanding. The parallels between hermeneutics and "clinical" methods of inquiry are very obvious to me: ... the hermeneutical demand for the researcher to reflect himself "into" the text he is producing. A social scientist is himself part of the society he investigates, so his own position must be taken into account. To me, then, the ... course is a way of making sense of this demand, giving it a concrete content. The hermeneuticists reached their views by philosophical analysis, and it is not

> incumbent upon them to give any cues as to how this "reflection of the self into" could possibly be accomplished. The ... course is one way of accomplishing this. It is mentioned as important in every book on qualitative research methods I've ever read, by the way, but these books, too, are silent on how to accomplish it.

In writing, "I allow certain themes, paragraphs, statements or propositions to capture my interests," this person acknowledges the interchange between the individual and the environment. She allows figures to emerge – what comes into figure is by definition of interest to her; otherwise it would stay in the background. She mentions that her students are worried about learning the "expected thing." Paradoxically, this concern, even though it may not always be in one's awareness, will interfere with the natural process of figure formation. If she can help her students put their worries to rest, what a profound metalearning experience her students will enjoy as they read their philosophical texts.

The process awareness training is an exercise in seeing and hearing what is actual and present. There is no "expected thing" to search for. The "investigator" becomes open to what is, allowing a fluid figure formation. This experience, which sounds extremely simple, is completely new for most people and not easy. Students often worry about doing the "right" thing, or doing it the "right" way. They tend to make judgments about what is important rather than allow the significance of a happening to present itself; they make interpretations, give reasons, and in other ways interfere with what simply is. Along with the stated skills, the training provides the experience of trusting process, teaching that, in all circumstances, something will emerge into figure.

An explicit goal of both the formalized process awareness training and process awareness work (chair work) is to increase awareness, but to everything we do in the course we apply awareness questions to remind the students constantly to shuttle their awareness between themselves and others. The following letter speaks of the consequences of the researcher's (not unusual) needs to be liked and to be helpful:

> I have attended the course twice.... Even if the second course also was a beginners' course that wasn't meant to build on the first one, it was somehow like a second-level course for me. The intuitive understanding of what awareness means in different situations comes gradually and only through a lot of training. I felt that I got new insights with every exercise we did, even if I had done it before.... In my work this affects my relations with the students in a positive way that enhances a two-way communication. My knowledge [of] how to be a researcher has definitely been altered by your course. The skills described above will of course influence my abilities as an interviewer too. The impact of my own personality on some of the choices I have made, working with my thesis, [became] very evident. Choices of subject, methods and literature as well as the interpretation of interviews have been influenced by my personal experiences, interests and preferences. This thought wasn't new. But I have discovered that not all the choices were so rational as I liked to think. The experiences in the course helped me to see some more irrational factors, too.
>
> Analyzing my interviews, I was puzzled discovering that there was a lot of good information in an interview which I had been discontented with and had considered as an interview with little relevant information. Reflecting over my personal involvement in that particular interview in terms of "what happened in that situation, how did I feel about it?" made it suddenly clear that my reason for classifying that interview as "bad" was not its content, but the feeling that the informant didn't like me! Another example is the need of being considered as a helpful, empathic person that sometimes made me forget the more scientific purpose of my research, making me focus too much on my informants' suffering instead of other information that would help to understand the phenomena central to my research theme.... Many books about research methods underline the importance of [minimizing] the effect of the researcher's personality on the research process. The Gestalt method gives the researcher a concrete tool ... to reach this goal.

This letter explains the advantages of a second course. Another student said, "When I took the course the first time you gave it, I was so immersed in my own material it didn't sink in. I didn't digest it." Within a few days, most students overcome their initial concerns about the novelty of the course and their lack of acquaintance with the other members of the class, but by then the course is a third or half over. By taking the course a second time, a student can continue the process that is only begun in the ten days we meet together, and increase his or her chances of making our approach operational. The student quoted above had already conducted her research interviews before her first course with us. Now she has learned some concrete tools for "how to be a researcher."

The following letter strikingly illustrates how students are able to continue building on what is begun in class:

> After I had finished the course I started my interview-study here in Oslo.... We have done about 50 interviews now. In our interviews we try to get a picture of how it is to be a doctor or specially trained ambulance crew called to an unconscious person, where you as a rescuer within minutes have to decide if you should resuscitate the patient or not. We also ask them questions about how it really is to "decide life over death" and how it is to live and work with acute medicine out of hospital. It is very rare that the health workers in Norway talk about the strain it is to have responsibility for other people's lives.
>
> We have tried to avoid "making them" answer what they expect we want to hear, and I think it has been of great value to me that I all the time tried to view myself from the metaposition during the interviews. I am (thanks to the things I learned at the course) all the time very aware of my own feelings and reactions to the things said and done by the interviewed person (and hopefully of myself as well). And I tried to use "the non-verbal communication" with the subject of the interview. (Example: "I can hear that your mouth is very dry now ..." / "Your voice is so low/sad when you tell me this ...") I could also tell them what it did to me to listen to them and to go into the things really deep together with them. This helped me build a kind of confidence/honesty in the interview situation; I think it made me less dangerous as an interviewer. I was open and honest so they could be open and honest. If we got into "difficult" situations I could solve the problem because I had learned how to put words [to] what I felt and thought and what I saw....
>
> As a result of this experience I have applied to the medical faculty asking if they can accept this course as "counting" in my research education. Personally I think it must be a perfect and very useful course for anyone who is doing qualitative research, interviews, etc.
>
> There is, however, another area where the course has been of great value to me.... Due to the work I do as a researcher I repeatedly find myself in situations with relatives who have just lost a beloved one. These situations always come "unprepared" and are very delicate, and it has always been a challenge to find the right things to say and do. How do I communicate [in] these situations where I as a doctor have limited time to give comfort, explain the medical problems and give actual information about the post-mortem, the undertaker, etc.? Most important,

> maybe – how do you get on the "wavelength" of a grieving person? This is maybe one area where I felt that I really could use what I learned during the two weeks last fall. I can tell the relatives that I see their reaction and how I feel. I can tell them what it is like for me to come into their life in a situation like the one they're in (whatever situation that might be). It is so much easier for me to establish contact and to make them feel safe and taken seriously.
>
> As doctors you are trained [to pay] attention to the "signals" given by the patient. But no one ever told us that it could be useful to put words to those signals in the communication with the patient nor the importance of the awareness of your own self in that communication. I think it would be a great advantage to have a course like this in the clinical education of doctors in Norway....

This researcher gives vivid descriptions of ways she applies basic person-to-person contact skills from the course. Simply being aware is not enough. She sets an example of self-disclosure, putting words to her responses to people. She makes herself "less dangerous" as she asks questions that may seem quite intrusive and could be interpreted as judgmental; she also brings the dialogue to a deeper level of honesty, increasing the richness of the data. She expresses what she notices about those she deals with: "Your voice is so low/sad when you tell me this." When we teach giving personal response and "stating the obvious," it is difficult to convey the potential effects of using these skills sensitively and without manipulation. This woman's willingness to make herself known appears to have had immediate positive effects on those fortunate enough to be in contact with her. By being in the moment, aware and responsive, she invites and gives tacit permission for the same from those around her. Together they join in a moment of shared reality and understanding.

This researcher mentions "all the time" directing her awareness out toward others *and* inward to herself. She is present at the level of interaction *and* takes a metaposition. This ability to shift one's focus rapidly is one of the skills we practice in class. The more it is done the easier it becomes until it is integrated into one's way of being. The potential of the attitude shown by this doctor is what Heshusius calls "merging": a way of being that is a "completely attentive, vigilant, but not intrusive turning of attention." It is "inactive activity" involving "somatic acts, acts of dwelling and indwelling, of extending oneself nonverbally, of letting go of self and attending a larger reality" (Heshusius 1994:18).

The writer of the above letter wants to find out "how it is to be [someone] called to an unconscious person where you as a rescuer ... have to decide if you should resuscitate the patient or not." Another student explores the experiences of innovative teachers; another, certain aspects of the immigrant experience. Such inquiries into the complexities of human experience require an orientation that allows the investigator to explore and analyze phenomena that can be neither counted nor measured, and then to put the raw data into a form that "can contribute to generating and/or confirming social science theory" (Patton 1990:94). The letters we received indicate that the course was beneficial in a variety of ways, including

- giving structure to journal writing;
- fostering the conscious use of oneself to improve the interviewing process and data collected;
- furthering the ability to take a metaview to increase the validity of the data;
- fostering an appreciation of the subjectivity of all experience;
- enhancing the understanding of the process of change;
- expanding ways of making contact and improving communication; and
- enriching the theoretical base of the students' research studies.

The practices and procedures we introduce in the course are not discrete bits of knowledge to be "learned" in two weeks. Rather, they are elements of a unified approach intended to enlarge the students' appreciation of 1) their own and others' complexity, 2) the mutuality of an individual and her or his environment – including other people, and 3) a process orientation in an ever-changing world. It is impossible to isolate a single classroom experience as *the* cause for a particular effect on a student's behavior or ability to carry out more stringent research. There is an overall consistency of basic principles to what we do, how we are, and how we encourage the students to be in class. The net result is a totality of experience, with each aspect supporting and reinforcing all other parts of the training.

The class interactions and experiments have serious intent; at the same time many of them invite playfulness, humor, flexibility, and creativity. The atmosphere encourages total personal involvement. The resulting expansion of awareness seems to be accompanied by a broader acceptance of diverse outlooks as well as an openness to, and tolerance of, novel experience and ambiguity. New ways of conceptualizing experience and viewing behavior and interaction become possible. The researcher is seen not as an "objective" gatherer of data, but as an active observer of relations, including herself or himself as a participator in the drama and an important source of data.

The course introduces students to a holistic approach to conceptualizing behavior and interaction. How we teach and what we teach together generate a different frame in which to carry out research. The course itself is a process of inquiry in which focus on the "now" and awareness are building blocks. These student comments, which came from their letters, reflect some reactions to awareness work:

- "I had some new experiences about how I influence persons around me."
- "The course made these dimensions of my personality more clear."
- "I frequently use the see/imagine distinction.... It's extremely facilitative."
- "I frequently detect how I come to conclusions about people."
- "Another realm where this course has proved extremely valuable to me is in the evaluation of the adequacy of (some) educational–philosophical arguments and pieces of historical research."

The purpose of the course "The Role of Self in Qualitative Research" is to improve the quality of research conducted by the students at the Institute for Educational Research at the University of Oslo. How can we know whether this actually comes about? The students recounted in their letters how they noticed their willingness to change perspectives of themselves and others and how they expanded their possibilities in conceptualizing and carrying out their research. They seem to experience or acquire increased awareness and flexibility; a focus on the "now"; openness to experience; and the ability to cope with complexity, change, ambiguity, redesigning systems, and reorganizing them to higher levels of complexity. They apply the awareness skills to their interviews and indicate they can more lucidly analyze and interpret their data.

Whether the 20 percent of the students who did not answer the letter had negative responses to the class we can only surmise. Also, it is possible that the excitement of the special and constant interactions, in addition to the unique atmosphere of the class, biased some responses. Nevertheless, we can conclude that at least some students made good use of specific learnings relevant to the qualitative research role. In addition, we believe that certain qualities, indeterminate and non-apparent yet considered necessary for human contact, were encouraged and nourished.

We hope others will explore ways to utilize subjectivity to promote more effective qualitative research. Perhaps readers who teach at the undergraduate and graduate levels will be inspired to reexamine the preparation of students for conducting research. We trust you will find at least some of the ideas we have put forth intriguing and worthy of being incorporated into existing training programs.

Those of you who are graduate students and anticipate doing research will perhaps seek out self-awareness and growth experiences that will challenge you in new and exciting ways as you begin your research projects.

With regard to readers who are in positions to influence graduate educational policy, we encourage you to seriously consider incorporating non-cognitive, self-focused learning approaches in your training, not only for those who will be using qualitative methodology, but also for those using other methods of inquiry. Of all the components implicated in conducting high-quality research, the self of the researcher is the foundation on which all else rests.

Chapter 8

Reflections and Recapitulations

To explain the observed phenomena we always *have to consider the wider context of the learning experiment, and* every *transaction between persons is a context of learning.*

G. Bateson (1972:246)

It has taken many pages to chronicle our course, "The Role of Self in Qualitative Research": placing it first in a historical context; linking it with the actual researcher as a human being; supporting it with epistemological, ontological, and theoretical background; and describing the actual procedures and processes. Now a backward glance and some comments from the metaposition are in order to bring a degree of closure.

Students enroll in this course knowing little about what is in store, and we conduct it in a language that is not the students' mother tongue. Furthermore, we can accomplish only so much in ten days. Nevertheless, students do become engaged in their learning – both cognitively and affectively – and, from their reports, appear to benefit from it.

Researchers usually use some kind of tool, or "instrument," such as a measuring device, a questionnaire, or a specially designed test. An instrument is generally thought of as a way of accomplishing a task by compensating for human limitations of strength, sensory capability, or mental or physical dexterity; it can be a means of extending or enhancing eyesight (e.g. a telescope), voice (a microphone), reach (a ladder), strength (a bulldozer), accuracy of measurement (mathematics), and so on. An instrument needs to be guided, manipulated, and applied to a suitable task. With its built-in mechanisms, it works by fixed rules and is directed by an intelligence outside itself; some person needs to understand its nature, vulnerabilities, limits, and what makes it work. An instrument is neither complicated by emotions nor distracted by memories. Referring to the researcher as her or his own instrument, as Patton, Berg and Smith, and many others have done, puts a great burden on the researcher – he or she becomes both the subject and the object, the user and the tool. Since the manipulator and the manipulated are one and the same entity, there can be confusion about what force is in charge: the instrument or the user of the instrument? Or, can they be separated, even for descriptive purposes?

Emotions in the classroom

It is no secret that humans have internal contradictions. Ibsen wrote that life is an endless war between the trolls in one's heart and brain. The human brain has a center for emotions, the ancient limbic system, as well as a thinking center, the neocortex, a later development. Our emotions are ever-present and can be incredibly overpowering, strong enough to distort or conceal an obvious reality when that reality poses a threat to our needs or desires. In this course we seek to familiarize our students with these basic complexities of human subjectivity. It is an ambitious goal and a formidable challenge.

The usual academic course is concerned primarily with intellectual activity for which there are prerequisites, some assumed and others officially stated; such a course demands readiness and some level of competence regarding the subject to be studied. This course requires a readiness that is different but equally legitimate and justifiable: in addition to intellectual skills, emotional and social resourcefulness are necessary. Students who join the course have no designated preparation and little conception of what they are facing, although the course announcement (Appendix A) mentions "training experiences ... based on

Gestalt approaches and group process" and that "the class is intensive and demanding." Only when students are participating in the day-to-day class activities is it possible for them to know whether they feel safe with the actions and interactions that are part of self-scrutiny.

This sobering reality could be used as an excuse to avoid or dismiss experiential activity related to emotional and social knowledge as inappropriate, non-academic, or incompatible with the existing paradigm. However, the emotions involved in self-exploration constitute an essential aspect of our humanness; more to the point in this work, self-exploration is a fitting preparation for all research endeavors and a prerequisite for using qualitative methods. It is the students' responsibility at each moment in the class to choose the level of their engagement. It is the responsibility of the instructors to be constantly aware of what is going on with the individual students and to have the sensitivity and experience to deal positively and constructively with whatever emerges.

The qualitative–Gestalt connection

My becoming a catalyst in this process for students and researchers has been one of life's surprise gifts. I had not anticipated being in a situation where my years of experience with the Gestalt approach would be so applicable and adaptable in quite a different context. However, I might have had a clue some years before; I first recognized the similarity of Gestalt principles to those of qualitative methods during a course I attended at the University of California in Santa Barbara. As the instructor, Dr. Laurence Iannaccone, described the process orientation of qualitative research, I noted some of his comments:

- As an interviewer, ask short, clear questions.
- Allow interviewees to bring up topics and words salient to them.
- Questions are the stimulus for response from the interviewee.
- It is from the response of the interviewee that the interviewer discovers her or his next intervention.

- Do not shape the interview any more than you have to.
- Do not control. Be controlled by the data.
- All steps in the process are mutually dependent.
- You do not have to know your target at the beginning.

I was pleased to feel so familiar with this kind of interviewing, this method of working with data: using thick description, separating description from inferences, allowing categories, themes, and patterns to emerge. All the above notions are analogous to a Gestalt interchange. But it was the focus on holism and on process – the actual mention of the importance of present-centeredness for the interviewer – that clinched it for me; it was then I settled into a feeling of comfort; I was no stranger in this territory. The connection I perceived between the two disciplines was supported by remarks I heard time and again at a conference of graduates of the confluent education program – people working in research said the Gestalt training they had received in the program was invaluable to them. In Chapter 2, I cited the researcher who found it impossible *not* to involve herself, even though the design of the research demanded that she adhere to a given set of questions. She told of another research project, in which she was an analyst of data:

> There was a qualitative difference between someone who had Gestalt as a background and someone who did not. One person, recognizing when a participant really wanted to say something more by their facial movements, picked up on it. He'd say things like, "You're nodding," and, "Well do you want to say anything about that?" And then the interviewee would go into a whole long discussion about that topic, whereas the other interviewer, not one time did he pick up on any kind of nonverbal cues. In fact, he even shut people down when they were giving him cues. So he was not aware at all of material that was beneath the surface. And he wasn't looking for anything in depth.

Each time we teach the course, I rediscover the value to students of the connection between the Gestalt approach and qualitative methodology. Now as I review my manuscript, I notice again how the two approaches blend. I feel privileged to straddle the two disciplines and serve as a link between them, to carry knowledge and experience to enrich a different milieu. This satisfaction is like that of finding in your closet just the item – an article of sports equipment, a tool, a garment, or whatever – that answers your present need (with perhaps a minor adjustment). It was this reality[1] that originally took us to the University of Oslo and subsequently inspired this book.

The amazing fit of the Gestalt approach and qualitative methodology impresses me each time I read Patton's "Ten Themes of Qualitative Inquiry" (see Chapter 3) and Lincoln and Guba's "Five Axioms of the Naturalistic Paradigm" (see Chapter 3). As I reread my manuscript, I am moved to open Lincoln and Guba and take another look at the "Characteristics" that follow the "Five Axioms." The following quotation, taken from "Characteristic 1," again illustrates the striking correspondence between Gestalt and qualitative theory:

> Naturalistic ontology suggests that realities are wholes that cannot be understood in isolation from their contexts, nor can they be fragmented for separate study of the parts (the whole is more than the sum of the parts); because the belief that the very act of observation influences what is seen, and so the research interaction should take place with the entity-in-context for fullest understanding;... because of the belief in mutual shaping rather than linear causation, which suggests that the phenomenon must be studied in its full-scale influence (force) field; and because contextual value structures are at least partly determinative of what will be found. (p. 39)

A metaview of the class

Now, as I review what I have written, I remind myself, don't explain, don't start writing another book – take a metaview. So to make it easier, I imagine for a moment that I am floating up near the ceiling of our classroom and looking down. I imagine us (George and I) instructing down below, wearing T-shirts printed with maps showing areas of interest that we synthesize in designing the course and providing a theoretical foundation for it. I see a variety of regions labeled education, human behavior, philosophy, sociology, creativity, communication, change, family therapy, organizational dynamics, and literature, as well as a large sector labeled Gestalt Practitioner and Trainer.

I can see from this vantage point that we keep the system open. Our plan is flexible enough that we can respond to emerging needs of individuals and the small groups. We make alterations in the schedule, adding or deleting experiences as time dictates. We take time to listen intently and without judgment to students' comments and questions, and to reply. This is what we teach, what we model, and what we hope they will take with them and apply in their research process. At the same time we cover the major components of our agenda. I greatly enjoy seeing how we use the students' questions and remarks as a springboard for our creativity, such as when we make up an exercise on the spur of the moment so the students make some kind of discovery through experience rather than through a lecture presentation.

I observe, from my imaginary position near the ceiling, how we engage the students. Each one participates. They talk together and laugh and cry at times. I am gratified to see how earnestly they invest themselves in the exercises and how involved they become in taking the roles of interviewer, participator, and observer. They behave in ways that would look peculiar to an uninitiated onlooker. I hear the word "now" spoken frequently. And what is this? They stand and talk to empty chairs! Now, that's new! What can they possibly learn from that? All these activities are interwoven with metaprocessing, the reading assignment groups, and discussions. "What are they learning, with all this engagement?" I ask.

Now I come down from my metaposition and reflect on the theoretical foundations implicit in our two weeks of activities. What I have observed is how the psychology, philosophy, and attitude of the Gestalt approach manifest themselves when translated into education – when we make what is implicit explicit. Students mobilize energy, seek and make contact with themselves and others, create events in the moment, and come to trust themselves, others, and the process they together allow to happen. When the process leads to closure, they discover the feeling of satisfaction it brings. And if and when they are left in an unfinished situation, they experience the effect and the meaning this has for them in their lives and research.

I realize as I review what I have written that there is almost nothing about gender issues. Most of our students are women. Some days George has been the only male in class, and on one of those days a student pointed this out. Informal discussions have revealed that, for some of our students, gender is an important issue; why is it, then, that the topic almost never arises? I will pay attention to this in the future. I think now how important it is to deal with all stereotypes and prejudices as they arise.

The students

Before each new course we wonder, will students come? Can they know what they may find useful when they do not know what is possible? I remember the comment of one student: "I wrote a book on experiential learning and I didn't know what an experience was!" If we do not know what we need to learn, how can we go looking for it? Yet students do arrive, in limited numbers. Each year the word spreads to friends and colleagues of those who have previously attended. From the start there has been discussion concerning the possibility of making the course a requirement for doctoral students at the Institute for Educational Research. Our reaction to this suggestion has been negative. The success and value of the course depend on the participants' openness to venture into new experiences, to be stimulated both intellectually and emotionally. If students were told they must attend, there would be a greater chance of polarization: on one side, those students who are glad to participate, eager to jump in with both feet; on the other side, those who resent taking two weeks away from their usual activities. They may already feel satisfied with their research, level of awareness, and interactional skills. Were the course of longer duration, the reactions of such students could provide a fruitful opportunity for dealing with issues commonly met in academic and research situations, such as required courses. In a two-week course, however, the presence of students who do not want to be there could interfere with the primary agenda and have considerable adverse influence on the mind-set of others. Even among those who come willingly, there are always some who are hesitant and not sure they are in the right place. Undoubtedly, the disadvantage of not making the course mandatory is that others who might benefit miss the experience and consequent learning.

One student who participated in the first course has continued at our invitation to attend and serve as an assistant in each successive course. Kari (not her real name) is a doctoral student at the institute, writing her dissertation in philosophy of science and teaching courses. Her comments, both in class and in metaprocessing sessions at lunch or after class, help us and the other students, who find her interesting and informative. Because her background includes knowledge and concepts outside our fields, everyone listens carefully to her remarks. Since she too is a student, the other members of the class go to her with observations and problems they might not share with us. She takes part in the exercises if an extra person is needed to make a group of three or a pair.

For the next course we have invited a few former students to come back – those who might later train researchers or incorporate aspects of the course in teacher training. (Some of the doctoral students are also on the faculties of teacher training and nurses' training institutions.) Whether the students' interests lie in research or training, or both, they have expressed a desire to deepen their understanding of our methods and goals and to receive more practice in the skills we teach. They will spend additional time at the close of each afternoon with Kari and us to review the pedagogical strategies of the training.

Having reread Chapters 5 and 6, and keeping in mind the returning students who might incorporate aspects of the course in their teaching, I must restate that this is not a recipe book. Rather, it is intended to be a description of a series of experiences presented during a two-week period to deal with the role of self in qualitative research. As described in Chapters 5 and 6, the course emphasizes the expansion of awareness, of personal knowledge, and of effective interaction and contact with others, always growing out of the philosophical roots of the Gestalt approach. I hope the material presented here will provoke interest, stimulate creativity, and inspire the development of more programs to answer the need for such training.

The lived adventure

Near the end of every course, at some point during a group metaprocess, a student has remarked, "I never would have enrolled if I had known what this course was like, but I am so glad I did." Other students nod their heads at this, smile a bit, and verbally agree. Another typical remark: "Although I read the description of the course before I enrolled, I didn't realize what it was." For me these comments reveal the huge gap that exists between the experience of the course and the words that describe it. The reader may also experience this gap.

This reminds me of what happens when telling a dream. I tell my dream and you listen, but it seems impossible to describe it such that you can experience the dream as I did. Suppose I begin this way: "I am walking down a narrow street in a very old village. The road goes down a hill; there are buildings on both sides of the road...." You may get a picture when I tell you this. But do you see *my* picture? What is more, no matter how many details I relate, is it possible to convey the feelings that accompanied my dream experience? If I say there was an eerie sense of impending doom, does your heart beat a little faster? Does your chest tighten and your gut contract? Feelings, and *talking about* feelings, are two very different modalities. Reading about a course is an avenue to understanding what the course is *about;* sitting in a room with other students and participating in events that tap into one's emotions is to experience actual emotions and feel sensations that are forthcoming. In one case we talk and think about, analyze, and categorize, using predominantly left brain activity; in the other, we undergo an experience and respond with affect, in which right brain activity is dominant. When these two situations are combined in an adventure, all systems are "go": the sensoric, the motoric, and the cerebral or mental.

Apparently, in much of life, we rehearse what we will do and say before we enter into what looms as a new and unpredictable experience. Shakespeare wrote, "All the world's a stage," and we of course want to keep to our script. Unchecked emotions, and the feelings that derive from them, are frequently thwarted. We stave them off; we do not want them "happen" – except for those considered acceptable. (Laughing is usually all right, crying is usually not.) Emotions are natural reactions to different stimuli. They are, in and of themselves, neither good nor bad. Yet to express emotion in front of others seems, in many cultures, wrong, dangerous, foolish, or all three. Thus, we feel more secure when we imagine the future and prepare for a happening before we walk into it. We lessen fear-laden anticipation by thinking, I'm ready; I know what I'm going to say and do; I won't be caught by surprise; I won't lose my temper or cry, or do anything embarrassing; I will present myself in the way I want other people to see me. We may not actually say these things to ourselves subvocally. Thoughts dart quickly, maybe so quickly and automatically that we do not even notice. Yet we guard ourselves against the unplanned, spontaneous event; we protect against authenticity. In our efforts to impress others, we literally *impress* instead of *express* ourselves; we show nothing, give nothing of ourselves away. Our hope is to achieve an outcome that fulfills what we need most from the interchange. Sometimes this simply means maintaining the image we want to present to the world – strong, secure, and in charge. Then we are all right.

In class we attempt to provide a place to experiment with new possibilities. We guide the students in two ways: in learning and forgetting.

Forgetting
- objective reality
- a "right" and a "wrong" way
- a desired outcome

Learning
- discovery
- to be surprised
- to surrender to the forming moment

Forgetting
- control and manipulation
- "why?"
- reasons, linear thinking, and figuring out

Learning
- unfolding process
- "what?" and "how?"
- "now" focus
- clear gestalt formation
- holistic relational view
- to make the implicit explicit
- to make the tacit manifest

I must stress that the instructors cannot achieve the learning and forgetting. That is the up to the students. Our job is to supply the situations, the atmosphere, the safety. We cannot know what any particular student will derive from a given experience. We may wonder, and even ask a student, "What are you holding onto in order to forbid a spontaneous moment?" Usually, no verbal reply is possible to such a question. What we can do, however, is foster awareness of the present moment, the student's process. The questions are basic: "What you are doing now? How are you doing it?" It is through awareness in and of the moment that new relationships are possible. I picture something that is difficult to capture in the telling: the moments of awakening in the students. The aha's. "Oh, this is what I am doing!" They have discovered a new way to organize their environment, and a novel figure emerges. "Aha!" The formation of a new gestalt.

We encourage experimentation in this safe, non-judgmental environment. These are the circumstances in which learning – unplanned and unexpected – takes place. Since emotions are not always predictable, there are surprises in unrehearsed moments: a burst of laughter, anger, or tears. Thus, for some students, the course holds an element of danger in the form of unanticipated, unrehearsed moments.

Added to such "hazards" is the unusual subject matter: the students themselves in connection with their roles as researchers. Have they ever been asked to experience the moment that is *now?* Or, to make explicit their internal monologue, or dialogue when they discuss two sides of an issue with themselves? Those are only two of the many challenges they face. On the first day of class we inform the students they do not have to do anything they do not want to, and they may leave the room at any time. The responsibility for what they do in class is placed squarely in the lap of each person. This, too, can be seen as a frightening predicament, since making choices can be difficult. If someone else gives the orders, we can sit back and comply or not. The risk level diminishes as our level of engagement lessens. Conversely, whatever we willingly assent to, we are at least partly accountable for; we cannot so easily blame others. In this course the students are aware from the very beginning, and perhaps before, that when they walk through the door they come to a place of adventure, risk, and discovery.

Responding to stimuli is natural for all living things. In most cultures today, our emotions are played on for responses – to buy a product, watch a film, vote for a certain candidate, and so on. Drama stimulates emotions and catharsis, an element of existence for which all societies have provided safe outlets. (In addition to drama, rituals and ritualistic healing also serve as outlets for strong experiences. See Scheff 1979.) Shakespeare's greatness was due in part to his genius in creating "aesthetic distance ... a balance of engagement and detachment" for the spectators (Scheff 1979:61). In our classes we provide a balance of risk and safety in connection with intense experiences. The unfamiliar exercises are "risky"; they are new, unrehearsed activities planned to evoke self-discovery and facilitate learning. Safety is provided by the support of the group, the setting of a university classroom, trusted instructors, and a non-judgmental classroom milieu. Again, we remind the participants that they do not have to do anything they do not want to do, and they have the freedom to reject any experience and/or to leave. The quickest route to safety, however, is inbred in each of us: When uncomfortable feelings surface, our immediate inclination is to ask ourselves, or another, *why*. The rapidity with which we shift to the cerebral mode is the ever-ready shield we all carry. *Why*, as it is usually asked, eclipses feelings and steers an interaction into a cerebral mode of explanation of feelings. It closes the door to the possibility of participatory consciousness.

Response in human interaction is of the moment, a feature of present-centeredness, openness, and acceptance. Mutual response is what creates relationship; it is the ground from which participatory consciousness comes into being. It is not only central to "self"; it is self-manifested.

Wheatley writes, "This world of relationships is rich and complex. Gregory Bateson (1980) speaks of 'the pattern that connects,' and urges that we stop teaching facts – the things of knowledge – and focus, instead, on relationships as the basis for all definitions" (1992:34).

"The Role of Self" course entails the whole person in relation to others. How could it be otherwise? We offer students the chance to move beyond habitual behaviors to whatever degree they choose. We invite them to step out of their familiar territory of habitual explanations, rationalizations, justifications, and routine, safe behaviors. A prime example is the process awareness training, in particular the "worst interviewer" experience. In this interaction, unlike anything one is accustomed to in interpersonal relations, people do give themselves over to a forming moment. As a rule the students experience much energy and laughter, and they surprise themselves with their creativity.

Perls's definition of learning – discovering that something is possible – embraces the idea of responding differently, taking a point of view one has not taken before, behaving other than our usual, habitual way. The relevance of the concept of change to the course is evident. Students not only learn through their experiences, but at a higher level, or metalevel, they learn they can participate in atypical ways. They acquire trust in their ability to respond to new situations, invaluable when doing qualitative inquiry. They discover it is possible to experience and express emotions. And they find that they are not frightened or repelled when others express their emotions, but instead are touched, pleased, and enlivened.

A major theme that emerges for me when rereading my manuscript is that of effectual change. Like one of a set of Russian dolls, change is nested within *learning* and *experience, feelings* and *emotion.* Thus, my reflections on change as an important element in the course, and a possible outcome of it, shift from one aspect to the other.

Full circle

The initial inspiration for this course came from a desire to change the education of doctoral students to make it possible for them to change and improve their way of conducting qualitative research, which necessitates a fundamental change in attitude and behavior. *Change* is neither inherently positive or negative; its meaning depends, as meaning always does, on the context. It is sometimes used in a clearly negative sense: "You've changed" (spoken as an accusation); or, "He's gone. Everything has changed" (spoken with grief). These examples express loss, and change always brings loss; we do not have what we had before. Change can seem like a threat if it is someone else's idea and we are not consulted during its planning and execution. "Change agents" believe they can facilitate change. A change agent may be a teacher, one who leads and encourages others to increase their options and improve their lives by discovering something is possible; but a change agent can also be a dictator who manipulates others in an oppressive way for their own needs and desires, usually with destructive consequences.

While I was rereading my manuscript, I also read "On Intellectual Craftsmanship" in *The Sociological Imagination* by C. Wright Mills (1959). Addressing the beginning student, Mills wrote, "you must learn to use your life experience in your intellectual work.... As a social scientist, you have to control this rather elaborate interplay [between life and work], to capture your experience and sort it out" (p. 196). I wondered about the apparent lack of self-involvement training for social scientists during the last thirty-five years. He instructed his readers to accomplish this "interplay" by keeping a journal: "the sociologist's need for systematic reflection demands it" (p. 196).

In the light of this early recommendation that social scientists use their life experiences – in other words, their selves – perhaps the absence of systematic preparation for accomplishing this illustrates resistance to change. Perhaps it has been simpler to avoid the issue. There are other possible explanations, to be sure. One is simply the lack of skills and conceptualizations necessary to effect Mills's ideas. I reflect on the expression "the time wasn't ripe," which is incontestable. If it had been ripe, graduate schools would have developed suitable courses, using affective as well as cognitive methods. Perhaps the alternative paradigms in several fields – education, philosophy, physics, psychology, astronomy, and others – signal a gradual shift. Despite narrowly focused specialization, there is some cross-fertilization, a coming together of ideas from many disciplines.

The groundwork necessary "to join personal experience and professional activities" (Mills 1959:196) includes, as we have interpreted it, instruction and practice in opening oneself to encounters that transcend one's everyday experience, expanding awareness of oneself and others, stimulating the interest and ability to take a metaview of situations. Along with these stated goals are additional beneficial effects: appreciation of differences, increased empathy, and the ability to foster and engage in moments of participatory consciousness. Having such a foundation can enrich meanings, reduce errors and distortions, and help to sort out and combine more effectively one's life experiences and scholarship. Mills, addressing those who would keep such a journal as he recommended, wrote, "you will not be afraid to use your experience and relate it directly to various work in progress.... The most admirable thinkers within the scholarly community ... do not split their work from their lives" (p. 195).

This statement brings us full circle to philosophers (early and more recent) who have not disassociated life and work. One might ponder the possibility that science is reassessing what it apparently lost back in the sixteenth century when radical changes in procedure and methods were instituted; the advent of the "scientific method" meant that human factors should be barred in the pursuit of objectivity, that life experience was to be kept separate and to claim no relevance in scientific work.

The issue that Mills raised in 1959 evidently has, until now, had little impact on the education of social researchers. This might be accounted for by the inertia of the system and general reluctance of educators to make changes, whatever the reasons. The issue raises provocative questions: What has actually happened in the last thirty-five years that is relevant to the issue raised by Mills? What kinds of approaches might be originated by others who are interested in training researchers to maximize the use of self? And in the training of trainers of researchers, as well? For those to be trained in the natural sciences, how relevant are the approaches contained in this book? Will the recent surge of interest in qualitative methods, manifested by doctoral studies and numerous books and articles, stimulate the development of teaching aimed at the affective domain as well as the cognitive? We have handles to grasp now: experiential learning, humanistic approaches, I–Thou conceptualizations, participatory consciousness, holistic medicine, and process orientation are examples. These can be more than mere abstractions when taken seriously and translated into training and practice.

In his *Self Portrait,* poet Patrick Kavanagh wrote, "My ancestors lived in the cave of the unconscious and screamed when they saw the light." He escaped the cave. He took a metaview and welcomed the light.

A final reflection on a core issue raised in Chapter 1: Scientific objectivity turns out to have a paradoxical twist. It could well be that the more human factors are ignored in doing research, the less true objectivity can be achieved. Conversely, the more the scientist increases awareness of herself or himself as a human being and makes use of this, the more the approximation of "virtual objectivity" will be enhanced, along with the minimizing of error.

The importance of the I in SCIENCE goes far beyond correct spelling.

Appendix A

[The following is the announcement from the University of Oslo to publicize the course.]

COURSE: THE ROLE OF SELF IN QUALITATIVE RESEARCH

INSTRUCTORS: George I. Brown
Judith R. Brown

GOAL: To improve the quality of one's research in the social sciences through emphasizing awareness of the "self as researcher" in the inquiry process.

RATIONALE: Awareness of self is often diminished as a result of socialization processes. We learn to look repeatedly to others, e.g. parents and teachers, for information as to what is happening and how this material is to be assessed. Rather than attend to data provided by our own experience we remain out of touch with ourselves, our emotions, feelings, desires, fear, etc. The resulting lack of trust in one's own experience interferes with using one's self consciously and intentionally as an instrument for obtaining and analyzing data in the research context.

All research occurs in the context of relationship. One can no longer assume a dichotomy between the observer and those being observed. Thus the researcher's awareness of self becomes essential for him or her to get accurate data from these relationships. The unconscious impact of our biases, beliefs and values on our perceptions and actions may distort, or even deny, the data immanent in relationships that occur in qualitative inquiry. Consequently, the demand for "scientific objectivity," for affective distance from others or from oneself, can instead interfere with the validity of the research.

THE COURSE: A sequence of training experiences will be provided based on Gestalt approaches and group process which will be supported by theoretical foundations. Both the experiential and theoretical are placed in the context of a researcher's function in qualitative methodology.

In order to become a more effective researcher some of what one needs includes:

- expanded awareness of oneself and others;
- appreciation of differences between self and others;
- increased ability for contact with self and others;
- competence in direct communication and self-expression.

Examples of questions used to increase self-awareness are:

What am I experiencing now? We discriminate between thinking, talking about, and experiencing, highlighting the role of emotions and self-expression.

What is happening now? What am I doing now? What is a metaview? "Meta" here means temporary emotional detachment from a situation in order to get a clearer grasp of ongoing interaction.

What is the difference between knowing and imagining? This question concerns connection and relationship, as do the following: How do I respond to you? How does my response affect you? How does one's existence threaten another's on an emotional level? How do I need you to be in order for me to be more comfortable?

How does motivation affect the awareness process? We will demonstrate and become familiar with the "Gestalt cycle of experience": the continual process of emerging needs, conscious and unconscious – often conflicting – and one's attempts to satisfy them.

FORMAT: The class is intensive and demanding. It will meet every day, Monday through Friday, May 3–7 and May 10–14, 9:15–12:15 and 13:30–16:30, and will be conducted in English. There are reading and other assignments every day. No one should apply for admission to the course unless he or she can make an explicit commitment to attend every day's complete sessions, both morning and afternoon. It would be extremely difficult to take this class and have other professional obligations at the same time, so applicants should plan accordingly.

The class will be limited to 15 participants. Doctoral students at the Institute of Pedagogical Research will be given priority but there should be some room for other students at the University and those from other institutions.

Because this course is unusual in both format and content, excerpts from evaluations of two students who participated in the first course offered are included here with their permission.

> I only wish that I had taken this course before I started my research.... In the first part of the course the personal experiences were dominating. Some of the exercises brought me in a powerful way in contact with my own feelings and with other people's feelings. We are not used to this, especially not in a doctoral course. But still they are there, all these feelings. And they shape our perspectives of the world and our interpretations, our choices and our relationships.... These new exciting experiences would have dominated if we hadn't our reading and reading discussion. They helped to see the connection to my research, and I can see how all I have learned will affect my analyses and the interviews in a fruitful way.
>
> Therese Sand

> In the beginning much of what went on was rather painful, in the sense that old "material" came much closer to the surface than it ordinarily is.

Somehow, then, the pain could be endured and even turned into something productive – that basically means learning something new to me. I don't know exactly how you accomplished that but one of the factors seems to me to be the constant pointing out of relations between the theory (Berg/Smith) and what we were doing in the class. That gave me a much better understanding of what happened in the class and at the same time provided the theory with substance that really gave meaning to what the authors said.

I have begun to think (the idea has actually crossed my mind before, too!) that philosophers of science cannot completely grasp what is going on in theory development unless they have felt the basic processes involved in data gathering on their bodies.

So for me this was extremely valuable in terms of the thesis I'm writing....

Tone Kvernbekk

TEXT: *The Self in Social Inquiry*, David N. Berg and Kenwyn K. Smith. Sage Publications, 1988.

Appendix B

Description of course: An introduction to theory, examination and practice of constructs of:

- awareness
- the Now
- seeing and hearing
- projection, introjection, retroflection, confluence
- presence
- contact – withdrawal
- shuttling

Objectives: to utilize subjectivity intentionally and consciously in research settings.

Procedures:
- lecture
- demonstration
- mini-lab participation in Process Awareness Training

Content:
- theoretical introduction
- demonstration through Process Awareness Work
- mirroring – to increase sight perception
- gibberish – to increase hearing
- separating seeing and imaging

- stating the obvious
- personal response
- hypothesis generation
- projection, introjection, retroflection, confluence as "mechanisms," interrelationships, healthy and pathological manifestations
- basic questions
- figure/ground
- relationships with existential Now
- metacommunication
- assuming a metaperspective
- reframing (generating alternative observational points)
- difference between description and judgments
- combining constructs in research functions
- contact – withdrawal
- orms of shuttling
- theoretical commentary related to the above
- discussion of ethical issues

Appendix C

September 5, 1993

Dear ....

As you may know I am writing a book about the course you attended with George and me here at the University of Oslo. This book emphasizes the importance of the self of the researcher and is intended for people who teach and/or use qualitative analysis.

I would appreciate information concerning how this course has been of use to you in your own work that I could include in the book. The more detailed your comments, the better (thick description).

If you have found the course useful in other areas of your life, that information would also be of interest to me.

We will receive mail here at the University.

Pedagogisk forskningsinstitutt
P.B. 1092, Blindern
0317 Oslo

Or: 2141 Ridge Lane
Santa Barbara, CA 93103
U.S.A.

I realize this will require some of your time; I thank you in advance for your help. Also, please mention if you have interest in a second level course.

I will be delighted to hear from you.

Sincerely,
(signed)
Judith Brown

Notes

Chapter 1

1. Quoted in Mahoney (1976:7).
2. In putting *the* before the word *self,* one conveys the idea of self as an entity, a static thing. This is contrary to the concept of self as a changing, developing process in evidence during one's contact and interaction with the environment. Self is also how one refers to who one is: "oneself," "I," distinct from anyone else; or as one's "person," the human counterpart of subjectivity, as in, "I use myself as an instrument in my work."
3. *Contact,* as generally used, to mean being in communication with, or making connection with, not necessarily with awareness. Used in the context of psychology, there is an understanding that awareness is necessary for contact.
4. Here, I create a word that for me incorporates substance, essence, and presence.
5. See Chapter 4.
6. This quotation is taken from an e-mail discussion. For further discussion of subjectivity, see Chapter 2.
7. This is not to suggest nothing happened in the Far East, or among the Arabs or the many other peoples who studied the heavens and the earth and made scientific discoveries.
8. For an in-depth discussion of this subject, see Berman (1981).
9. Berman (1981) refers to the separation of "fact and value."

Chapter 2

1. Margot Ely and her colleagues, in *Doing Qualitative Research: Circles within Circles,* use many excerpts from their students. These provide excellent illustrations of the students' thoughts and experiences as they engage in their research projects. Unlike Ely, I have omitted the students' names.

2. Hoshmand is referring specifically to the "reflective scientist practitioner."
3. People who are observed or interviewed or in other ways take part in a research study may be referred to as "participators" rather than "subjects" or "interviewees." When researchers have no privileged information, are not "experimenting" on subjects, are not in authoritative positions (they may even participate in the same activities as those being investigated), relationships tend to be egalitarian, with mutual trust and respect.
4. This excerpt is from a former student in the course that inspired this book: "The Role of Self in Qualitative Research," taught at the Institute for Educational Research at the University of Oslo, Norway.
5. See note 4.
6. Babcock (1980), Ruby (1980, 1982), Belmonte (1979), Berreman (1962), Crapanzano (1977, 1980), Dwyer (1982), Rabinow (1977, 1982), among others.
7. One's environment may also cause distortion in growth and development. See Sullivan (1953), Satir (1972), Bateson (1972), and J.R. Brown (1986).
8. See note 4.
9. From a personal conversation between Gregory Bateson and George Brown.
10. Lowman (1988:183) lists "personal attributes" necessary for the "clinical researcher": A. Cognitive complexity and high intelligence. B. Interpersonal competence, including 1. openness and honesty, 2. awareness of and control over one's own personal psychological defense, 3. cognitive flexibility, and 4. self-awareness and self-acceptance. C. Moderate needs for affiliation, achievement, and power. D. "Helping" orientation. E. High need for personal competence. F. Comfort with intuitive methods. G. Imagination and cognitive flexibility. See also Patton's "themes" in Chapter 3.
11. Participatory consciousness is not limited to person-to-person interaction. Berman (1981:16) states that "participating consciousness" involves "merger, or identification with" the natural world around us.
12. The goals of this education will be covered more fully in Chapter 3, under "Confluent education."

Chapter 3

1. This book has unnumbered pages.
2. Naturalistic inquiry "describes an alternative paradigm.... It has other aliases as well, for example: postpositivistic, ethnographic, phenomenological, subjective, case study, qualitative, hermeneutic, humanistic" (Lincoln and Guba 1985:7).

3. Since Perls's death in 1970, other practitioners have continued to enhance and modify Gestalt theory and practice.
4. Clarkson and Mackewn (1993:82) trace "the gestalt cycle," the process of self-regulation, to Perls (1947). It has been given various names: "the disturbance cycle," "the cycle of inter-dependency of organism and environment," and others. It was at The Gestalt Institute of Cleveland where "in particular Bill Warner first labelled it the `cycle of experience'" (Clarkson and Mackewn 1993:50).
5. In emphasizing the present-centered focus of the Gestalt approach, it is important to understand that the significance of the past and the future are recognized; however, all material is dealt with in the "now" experience.
6. These themes may all have meaning and usefulness in other methodologies.
7. For full explication of paradox, see Watzlawick et al. (1967).
8. A teacher was quoted as saying, "If I tell you, you will forget. If I show you, you may remember. If I involve you, you will understand."

Chapter 4

1. This differentiation between directed and undirected awareness is from de Vries (1993:25).
2. Appreciating the differences between oneself and another means differentiating between what is and is not "me." One is aware of, acknowledges, and respects the other as different, legitimate, and credible even though one may not actually like the other in all ways.
3. It seems that having rapport satisfies a deep need of humans to be in trusting and harmonious relationships. Familiar examples: we yawn, laugh, cry, and cheer with others.
4. The expressions "above the salt" and "below the salt" allude from the old custom of seating people of higher rank above the salt cellar placed near the middle of a long table.
5. This word, like the word *process,* can be used as both a noun and a verb.

Chapter 5

1. He attended the complete course when it was next offered.
2. Hoshmand mentioned the instructors' modeling behaviors deemed important for researchers: "use of self by teachers and supervisors and the modeling of reflexive questioning can facilitate similar reflective habits in students and trainees.... suspension of judgment and a willingness to consider what appear to be opposing possibilities or a breadth of meanings are intellectual habits and dialectical ways of thought that can be modeled" (1994:182).

3. Sullivan (1953), the “not me.”
4. In the Gestalt approach, *resent* is an important word for which there is no equivalent in Norwegian and some other languages. The Norwegian word *bebreide* is not the same. Resent means to feel angry or indignant toward someone about something. We speak of “harboring resentments,” meaning holding on to anger or indignation over a period of time.
5. Chapter 6 is devoted entirely to process awareness training.

Chapter 6

1. From Edwin Denby’s *Dancers, Buildings and People in the Streets.*
2. It has already been suggested that seeing is fraught with complexities. In Chapter 2, see Needleman's assertion that seeing is a “revolutionary act.” In Chapter 1, see Wheatley's contention that ours is “a participative universe” (Wheeler) where the observer is “part of the process that brings forth the manifestation of what we are observing.”
3. We are not against judgments or evaluations. It seems most of us are making them much of the time, whether we speak them or remain silent. In this course we emphasize descriptive feedback, which we think is important for several reasons. If students feel judgmental or are evaluating the behavior of others in the group, there is ample time for them to say so. They may do this during process and metaprocess sessions by being direct and expressing their thoughts.
4. In this context, *double-bind communication* refers to a verbal message that is negated by a non-verbal message: e.g. the statement “I feel relaxed and comfortable with you,” contradicted by a stiff smile and fidgeting hands. See Watzlawick et al. 1967:224ff.

Chapter 7

1. From “The Waking,” in *The Collected Poems of Theodore Roethke.*

Chapter 8

1. What is “reality” for me is based on my construction of reality. See Berger and Luckmann (1967). This “fit” that I perceive, which seems so evident and unmistakable to me, could possibly be construed otherwise by another.

References

Antonovsky A. 1994. A sociological critique of the "Well-Being" movement. *Advances: The Journal of Mind-Body Health* 10(3).

Babcock B. 1980. Reflexivity: definitions and discriminations. *Seminotica* 30(1–2).

Bateson G. 1972. *Steps to an Ecology of Mind.* New York: Ballantine Books.

Belmonte T. 1979. *The Broken Fountain.* New York: Columbia University Press.

Berg DN, Smith KK, eds. 1988. *The Self in Social Inquiry: Researching Methods.* Newbury Park, Calif.: Sage Publications.

Berger PL, Luckmann T. 1967. *The Social Construction of Reality: A Treatise in the Sociology of Knowledge.* Garden City, NY: Anchor Books/Doubleday & Company, Inc.

Berman M. 1981. *The Reenchantment of the World.* Ithaca, NY: Cornell University Press.

Berreman GD. 1962. *Behind Many Masks: Ethnography and Impression Management in a Himalayan Village.* Monograph No. 4. Lexington, Ky.: Society for Applied Anthropology.

Brown GI. 1990. *Human Teaching for Human Learning: An Introduction to Confluent Education.* Highland, NY: Gestalt Journal Press. (Originally published 1971.)

Brown JR. 1986. *I Only Want What's Best for You.* New York: St. Martin's Press.

Buber M. 1970. *I and Thou.* New York: Charles Scribner's Sons.

Clarkson P, Mackewn J. 1993. *Fritz Perls.* London: Sage Publications.

Cook T, Caston MA. 1991. *New Structures in Training: Exploring the Role of Systematic Processing in Achieving Relational Understanding.* Confluent Education Conference, Manhattan Beach, Calif.

Crapanzano V. 1977. On the writing of ethnography. *Dialectical Anthropology* 2(1).

Crapanzano V. 1980. *Tuhumi: Portrait of a Moroccan.* Chicago: University of

Chicago Press.

de Vries M. 1993. *Becoming Alive: The Counseling of People with Life-Threatening Disease at the Helen Dowling Institute.* Unpublished manuscript.

Dwyer K. 1982. *Moroccan Dialogues: Anthropology in Question.* Baltimore: Johns Hopkins University Press.

Ely M. 1991. *Doing Qualitative Research: Circles within Circles.* Bristol, Pa.: The Falmer Press.

Fisher H. 1994. Review of *The Challenge of Anthropology: Old Encounters and New Excursions,* by Robin Fox. *New York Times Book Review,* March 20.

Fox R. 1994. *The Challenge of Anthropology: Old Encounters and New Excursions.* New Brunswick, NJ: Transaction.

Gjertsen D. 1989. *Science and Philosophy: Past and Present.* London: Penguin Books.

Greenberg R. 1993. *Music Lectures.* Springfield, Va.: Teaching Co.

Heshusius L. 1994. Freeing ourselves from objectivity: managing subjectivity or turning toward a participatory mode of consciousness? *Educational Researcher* 23(3).

Holland J. 1973. *Making Vocational Choices.* Englewood Cliffs, NJ: Prentice-Hall.

Hoshmand L. 1994. *Orientation to Inquiry in a Reflective Professional Psychology.* Albany, NY: State University of New York Press.

Jourard SM. 1964. *The Transparent Self.* New York: D. Van Nostrand Co.

Kleinman S, Copp MA. 1993. *Emotions and Fieldwork.* Newbury Park, Calif.: Sage Publications.

Kleiveland J. 1992. *A Gestalt Approach to Organizational Change: Strategies and Interventions Used by an Internal Consultant, a Case Study.* Unpublished dissertation, University of California, Santa Barbara.

Koestler A. 1972. *The Case of the Midwife Toad.* New York: Random House.

Korb M, Gorrell J, Van De Riet V. 1989. *Gestalt Therapy: Practice and Theory.* Elmsford, NY: Pergamon Press.

Kuhn T. 1970. *The Structure of Scientific Revolutions.* 2nd ed. Chicago: University of Chicago Press.

Lems P. 1989. *The Executive Training Research Project: An Exploratory Study.* Unpublished dissertation, University of California, Santa Barbara.

Lichtenberg P. 1990. *Undoing the Clinch of Oppression.* New York: Peter Lang Publishing, Inc.

Lincoln YS, Guba EG. 1985. *Naturalistic Inquiry.* Newbury Park, Calif.: Sage Publications.

Lowman RL. 1988. What *is* clinical method? In Berg DN, Smith KK, eds.: *The Self in Social Inquiry: Research Methods.* Newbury Park, Calif.: Sage Publications.

McCall GJ, Simmons JL. 1969. *Issues in Participant Observation: A Text and Reader.* Reading, Mass.: Addison-Wesley Publishing Co.

Mahoney MJ. 1976. *Scientist as Subject: The Psychological Imperative.* Cambridge, Mass.: Ballinger Publishing Company.

Maslow AH. 1971. *The Farther Reaches of Human Nature.* New York: Viking Press.

Merry U, Brown G. 1987. *The Neurotic Behavior of Organizations.* New York: Gardner Press.

Miller DW. 1993. Reflections on the psychobiological nature of reality, with a theory about Alzheimer's disease. *Advances: The Journal of Mind-Body Health* 9(1).

Mills CW. 1959. *The Sociological Imagination.* London: Oxford University Press.

Mirvis P, Louis M. 1988. Self-full research: working through the self as instrument in organizational research. In Berg DN, Smith KK, eds.: *The Self in Social Inquiry: Research Methods.* Newbury Park, Calif.: Sage Publications.

Mitroff I. 1974. *The Subjective Side of Science.* New York: Elsevier Scientific Publishing Co.

Montuori A. 1992. Creativity, chaos, and self-renewal in human systems. *World Futures* 35.

Moon SM, Dillon DR, Sprenkle DH. 1990. Family therapy and qualitative research. *Journal of Marital and Family Therapy* 16(4).

Naranjo C. 1993. *Gestalt Therapy: The Attitude and Practice of an Atheoretical Experientialism.* Nevada City, Calif.: Gateways/IDHHB Publishing.

Naylor G. 1989. *Mamma Day.* New York: Vintage Books.

Needleman J. 1985. *The Way of the Physician.* San Francisco: Harper & Row.

Papernow PL. 1993. *Becoming a Stepfamily: Pattern of Development in Remarried Families.* San Francisco: Jossey-Bass.

Patton MQ. 1990. *Qualitative Evaluation and Research Methods.* Newbury Park, Calif.: Sage Publications.

Perls FS. 1947. *Ego, Hunger and Aggression.* New York: Random House. (Originally published 1942 in South Africa.)

Perls FS. 1969. *In and Out the Garbage Pail.* Lafayette, Calif.: Real People Press. (Pages unnumbered.)

Perls FS. 1973. *The Gestalt Approach and Eye Witness to Therapy.* Palo Alto, Calif.: Science and Behavior Books.

Perls FS. 1979. Planned psychotherapy. *The Gestalt Journal* II(2).

Perls FS. 1988. *Gestalt Therapy Verbatim.* Highland, NY: Gestalt Journal Press. (Originally published 1969.)

Rabinow P. 1977. *Reflections on Fieldwork in Morocco.* Berkeley, Calif.: University of California Press.

Rabinow P. 1982. Masked I go forward: reflections on the modern subject. In Ruby J, ed.: *A Crack in the Mirror: Reflective Perspectives in Anthropology.* Philadelphia: University of Pennsylvania Press, pp. 173–85.

Rice SA. 1929. Contagious bias in an interview. *American Journal of Sociology* 35.

Rosenthal R. 1966. *Experimenter Effects in Behavioral Research.* New York: Meredith Publishing Co.

Ruby J. 1980. Exposing yourself: reflexivity, film, and anthropology. *Semiotica* 30 (1–2).

Ruby J. 1982. *A Crack in the Mirror: Reflexive Perspectives in Anthropology.* Philadelphia: University of Pennsylvania Press.

Satir V. 1972. *Peoplemaking.* Palo Alto, Calif.: Science and Behavior Books, Inc.

Schachtel EG. 1959. *Metamorphosis. On the Development of Affect, Perception, Attention and Memory.* New York: Basic Books.

Scheff TJ. 1979. *Catharsis in Healing, Ritual, and Drama.* Berkeley and Los Angeles: University of California Press.

Schoen S. 1994. *Presence of Mind: Literary and Philosophical Roots of a Wise Psychotherapy.* Highland, NY: Gestalt Journal Press.

Simmons V. 1988. Reconstructing an organization's history: systematic distortion in retrospective data. In Berg DN, Smith KK, eds.: *The Self in Social Inquiry: Research Methods.* Newbury Park, Calif.: Sage Publications.

Smith KK, Berg DN. 1987. *Paradoxes of Group Life: Understanding Conflict, Paralysis, and Movement in Group Dynamics.* San Francisco: Jossey-Bass.

Spiro HM. 1986. *Doctors, Patients, and Placebos.* New Haven, Conn.: Yale University Press.

Sullivan HS. 1953. *Interpersonal Theory of Psychiatry.* New York: W. W. Norton.

Watzlawick P. 1990. *Münchhausen's Pigtail: Or Psychotherapy and "Reality." Essays and Lectures.* New York: W. W. Norton.

Watzlawick P, Bavelas JB, Jackson DD. 1967. *Pragmatics of Human Communication: A Study of Interactional Patterns, Pathologies, and Paradoxes.* New York: W. W. Norton.

Watzlawick P, Weakland JH, Fisch R. 1974. *Change: Principles of Problem Formation and Problem Resolution.* New York: W. W. Norton.

Wheatley MJ. 1992. *Leadership and the New Science: Learning about Organization from an Orderly Universe.* San Francisco: Berrett-Koehler Publishers.

Wiener CL. 1990. Introduction. In Wiener CL, Wysmans WM, eds.: *Grounded Theory in Medical Research: From Theory to Practice.* Amsterdam: Swets & Zeitlinger.

Zohar D. 1990. *The Quantum Self: Human Nature and Consciousness Defined by the New Physics.* New York: Quill/William Morrow.

Index